COLLABORATIVE RISK MITIGATION THROUGH CONSTRUCTION PLANNING AND SCHEDULING

COLLABORATIVE RISK MITIGATION THROUGH CONSTRUCTION PLANNING AND SCHEDULING: RISK DOESN'T HAVE TO BE A FOUR-LETTER WORD

BY

LANA KAY COBLE, ED.D, CPC
Tellepsen Builders, USA

emerald PUBLISHING

United Kingdom – North America – Japan – India – Malaysia – China

Emerald Publishing Limited
Howard House, Wagon Lane, Bingley BD16 1WA, UK

First edition 2019

British Library Cataloguing in Publication Data
A catalogue record for this book is available from the British Library

ISBN: 978-1-78743-148-5 (Print)
ISBN: 978-1-78743-147-8 (Online)
ISBN: 978-1-78743-248-2 (Epub)

INVESTOR IN PEOPLE

Contents

Synopsis from Industry Professionals

Owners

"Whether you're new to the industry or a seasoned professional; architect, project manager, contractor, or owner; and regardless the size of the project, this is a must read for all in the industry on how to 'plan your work and work your plan' for more successful outcomes."
Mark Webb
Principal at Vizient

"With the construction industry challenged with a limited labor market, lack of skilled trades, competitive landscape, and reduced margins, it is imperative that owners mitigate their risk in all facets of the capital project delivery process. With Lana Coble's thirty plus years of experience in the construction industry, serving as a general contractor, owners representative, faculty member, and architect, she brings unique and proven experiences, and perspectives in this book through case studies that predict risks, with proven solutions."

Spencer Moore, Vice President of Facilities Management and Operations
University of Texas M. D. Anderson Cancer Center

"Embarking on a large capital project is a challenging undertaking, even for a seasoned project leader. There is so much at stake — tens to hundreds of millions of dollars, the ability for an Institution to continue its mission and Institutional reputation, to name a just a few of the larger risks. It is not for the faint of heart or the inexperienced. Lana is neither of those, to say the least.

Lana has written a book, based on her long and extensive experience, to help both industry professionals as well as inexperienced owners to anticipate those risks and challenges. By doing so, they can better plan for and ultimately manage the risks that come from designing and constructing buildings.

Lana's depth of experience makes her extremely qualified to walk the reader through these challenges. She brings case-study experience to illuminate her concepts and assist the reader to better comprehend the ideas she wishes to share."

Sidney J Sanders, Senior Vice President
Construction, Facilities Design and Real Estate, Houston Methodist

Program Manager

"Set aside those 'how-to' project management books written by self-proclaimed authorities who've never managed a successful project. Make way for Lana Coble who in *Risk Doesn't Have to Be a Four-Letter Word* provides highly useful approaches, specific tools, and sage advice on reducing risks to greatly improve the chances for project success. As one of the most experienced, knowledgeable, and owner-oriented project leaders I've ever known, you can be assured her recommended techniques and scheduling practices have been tested in her 'laboratory' of real world projects. For me, she led very large, complex projects in Houston's dense and cramped Texas Medical Center and achieved outstanding results while gaining the respect of all who worked with her. Finally, a book has been written by a proven project authority based on doing, not on philosophy."
James A. Broaddus, PhD, P.E.
President, The Broaddus Companies

Architect

"This book will be a great resource for anyone who wants to understand the importance of integrating project planning (managing risks) with scheduling.

To most designers/architects' project planning and schedule creation are two separate and distinct exercises. Dr. Coble's book explains the difference between the two and how integral they are to a project's success. Congratulations on an informative and timely read!!"
Gus Blanco, AIA, ACHA
Studio Leader, Healthcare, Senior Principal
EYP Architecture & Engineering

Construction Agency

"While the commercial construction has made great strides in utilizing the latest technology and advancing less contentious procurement practices, delivering a high-quality construction project on time and within budget remains a complex proposition. In *Risk Doesn't Have to Be a Four-Letter Word* Dr. Coble presents in a very systematic way, strategies and implementation measures that go a long way in enhancing the likelihood of a successful project for all involved."
Jerry Nevlud, President/CEO
Associated General Contractors of America | Houston Chapter

General Contractors

"Risk Doesn't Have to Be a Four-Letter Word" Dr. Coble's passion for effective, team-oriented planning and scheduling shines through. The manuscript lays

out a great roadmap for those unfamiliar with effective project schedule management and provides a well-stocked tool box to build upon. Great Read!"
Thomas Kulick
General Superintendent Hensel Phelps

"I have worked with Lana Coble for 20 years and from Day 1 it was obvious that she possessed that innate ability to get things DONE! It has led to an association (and friendship) that benefits our clients and the construction community. Well done, Dr. Coble, I am looking forward to our collaboration on your next literary effort, 'The Art of Pre-Construction'."
Guy Cooke
Preconstruction Manager Tellepsen

"From personal experience, I know Dr. Cobles' energy, enthusiasm, and commitment to collaborative and effective project schedule management. For someone new to the industry, or a seasoned professional, this book is a must read for anyone eager to achieve positive project outcomes and relationships.

Dr. Cobles' unique qualifications and experience from the vantage point of Owners, Program Managers, Contractors, Architects, and Engineers alike provides credibility to the tools and processes outlined in this book. If you want your project to succeed, this is a must read."
Michael Dwight
Director of Operations Hensel Phelps

Subcontractor

"In a time when our industry is challenged with more complex projects being built ever faster, it is critical that we understand the risk. Lana has been one of our best thought-leaders for decades on better ways and methods to build the most complicated construction projects. In this book, she does an exceptional job of combining her experience and research to help us identify and minimize the risks that affect all members of the design and construction team."
Graham Moore
President TD Industries

Project Delivery Consultant

"Risk mitigation [...] engaging stakeholders [...] project planning vs. project scheduling [...] creating a culture of curiosity; all of these concepts are easily said but difficult to embrace both the art and science of their successful implementation for ultimate project success. Lana Coble is one of those rare technical talents who can explain those concepts but also specifically mentor folks in how to pull them off successfully [...] and in Lana's case, in an emotionally intelligent manner. Read this book several times. It's like a good novel, you will learn something new each time."
Wayne O'Neill
CEO RESET

About the Author

Dr. Lana Kay Coble CPC has dedicated her career to the advancement of construction methods, education of future and existing professionals, and promotion of diversity within the industry workforce. She has over 38 years of construction management experience, including positions as an owner's representative, executive manager in construction companies, and a professor in a major university construction management program. She has also published research in the American Institute of Constructors national journal. Her portfolio of construction projects includes healthcare, higher education, K-12 schools, commercial development, corporate campuses, mission critical data and energy facilities, athletic venues, zoo facilities, and major infrastructure systems. Her expertise includes the introduction of advanced technology to more efficiently engage all stakeholders in the construction process from concept to completion. Technologies and methods include implementation of software systems, development of document management systems, complex program management, planning/scheduling, and conflict mitigation on major high-profile construction projects. Ms. Coble's experience as an academician includes teaching at both the university level as well as in professional organizations and construction firms. Long a champion for diversity in the construction profession, Ms. Coble is an advocate for the development of women in construction and the advancement of historically underutilized businesses (HUB) through targeted affirmative action programs. Her passion for scheduling and mitigation is based upon academic fundamentals and construction experience. Within the span of her 38-year career, she has completed every project within the scheduled due date, with exception of one very complex project. Her most recent teaching endeavor was serving as the lead author of scheduling mitigation training to construction project managers and superintendents.

Dr. Coble holds a Bachelor of Arts in Construction & Design from Trinity University, a Master of Science in Architecture from Texas A&M University, and a Doctoral degree in Education from the University of Houston.

Purpose of This Book

To determine if this book has value for you, first and foremost it's important to understand what is not included in the pages to come. Specifically excluded are:

- instructions on how to utilize scheduling software,
- basic fundamentals and definitions of project scheduling, and
- historical reference to the development of critical path scheduling.

There are a multitude of publications which explain the techniques of schedule creation inclusive of logic, activity parameters, and critical path calculations. While it is beneficial to the reader to understand the fundamentals of project scheduling, this knowledge is not required to enhance schedule management. This book does not address how to implement software for schedule documentation, as Primavera and Microsoft Project primers are commonly available. While it is important to understand the origination of any process to be learned in that field, there are many textbooks which illuminate critical path scheduling. With that said, students, schedulers, and academicians may assume that this book holds little value, however the author believes the case is to the contrary. Concepts and practices listed below will demonstrate the value of this work for a breadth of construction management members ranging from students, schedulers, owners, architects, engineers, and contractors.

Content included in this manuscript is focused on real-world strategy and implementation measures to proactively avoid risk in the project schedule. Specifically included are:

- Understanding the difference between project planning vs. scheduling.
- Identifying processes which allow all key stakeholders to collaborate in the project planning phase by identifying potential risk.
- Defining a new conceptual perspective during schedule preparation that focuses on mitigating the potential risk elements before they occur.
- Forecasting the greatest areas of risk that can cause project failure and providing tangible tools to include preventive measures in the schedule.
- Providing practical application of risk mitigation strategies for commercial construction projects based upon case-study lessons learned from actual projects.
- Demonstrating periodic communication tools which allow understanding of the schedule status and risk identification in non-technical scheduling terms.
- Identifying the greatest areas of risk in actual production reporting to the scheduler which typically create late project completions, and

- Reframing the way that construction professionals and project team members think about the purpose of a schedule and its purpose during project implementation.

During the authors teaching tenure in both the higher education and practitioners classroom, the most common lack of understanding by students is the difference between planning and scheduling. Knowledge of the differentiators between both concepts is critically important when identifying who can participate in schedule preparation. Planning is for any stakeholder who can identify risk associated with execution of the project. The act of scheduling is reserved for those persons who are trained in software and understand each individual activity and the sequencing logic required to complete the scope of work. Observation has revealed that typically more time is spent on scheduling which is performed by the fewest number of persons. Common sense would suggest that greater benefit could occur by spending more time with a larger group of stakeholders in identifying which high risk activities should be tracked during project production. Suffice to say, both planning and scheduling are integral to project success, but planning is often overlooked by scheduling professionals. Hopefully, this text will bridge the communication gap between the members of the team which are not focused on production and those who are. Additionally, production personnel, that is, schedulers, and field supervisors, are not the decision-makers which emphasizes the importance that owners and key stakeholders understand the impact of risk and decisions on timely project completion. The framework presented in this book should also shape a new "way of thinking" by the construction practitioners during the development of a project schedule. By examining risk areas associated with the project, tasks can be assembled which would minimize these risks and placed in time appropriate slots of the schedule to allow anticipatory mitigation. This approach is not only proactive, but also communicates project aspects that the technical construction professional may not have been privy to prior to this risk identification process. A practical tool developed by the author, the "Risk Identification Matrix", can be used to facilitate this communication process by all stakeholders to enhance the development of the project schedule. Additional reporting tools are also included for the ongoing identification of risk as the project is constructed. After all, processes are dynamic and outside influences are always a risk during the production period. Schedules are "not wallpaper" and will continue to evolve as circumstances change relative to project conditions. This strategic shift in perspective on how to prepare and manage project scheduling should facilitate a more collaborative environment. Owners, program managers, architects, engineers, and construction professionals do not have to understand the technicalities of scheduling, rather they can learn how to transmit their concerns, so they can be included in the overall project schedule. Technical schedulers can benefit from learning how to discern key risk concerns by other team members and then translate those risks into schedule activities. While this book does not address the basic understanding of scheduling provided by academicians, it does provide the application framework for the content creation phase of development. Lastly, higher education students who do not have real project experience, will be provided a conceptual starting point for analyzing the project schedule based upon projected risk factors. The common beneficial thread for all team members will be a process intended to forecast risk factors for the purpose of inclusive planning for successful schedule management.

In writing this book, the author has taken into consideration that all the readers will be practitioners at some point in their careers, so the text contain summarized strategic points for quick reference. Being a construction practitioner for the last 38 years, the author understands how limited time is when executing a project and felt it is essential to provide key concepts which are easily identified. These tips have been time tested and are important to consider when building your team and providing direction for its members.

Acknowledgments

First and foremost, I want to thank all the industry practitioners who participated in the case studies for this manuscript. Your commitment to furthering the advancement of the building industry and the professionalism with which we practice is vitally important.

This book would not be possible without the tutelage of the academicians I had the privilege of working with during the pursuit of my doctorate degree. Profound appreciation to Dr. Lee Mountain whose belief in my message propelled me to pursue this project. I am eternally grateful for Dr. Cheryl Craig's ability to awaken the writing muse within me.

I would be remiss without acknowledging the many professionals who have contributed to my career as a building professional. Notably, Jay Tribble, who gave me my start as a construction project manager during a time when women project managers were an anomaly to the profession. A special thanks to the subcontractors who supported me and my projects for the last 38 years. To Donald Bonham and Jim Broaddus, thank you for the opportunities to experience project delivery from the perspective of an owner and program manager, respectively. And to Howard Tellepsen, my appreciation for providing me the opportunity to hone my skills as a general contractor. Profound appreciation to Brendan Jefferies and Guy Cooke who have been my building partners for the last 18 years of my career and for making construction fun!

Last and most especially not least, is my beloved Guusje who waited patiently for the last seven years while I sequentially completed my doctoral studies and wrote this manuscript. Your patience and support while I pursued my passion for education and giving back to the industry was truly a gift.

Introduction

The approach of this text is to present "real-world" construction schedule experience with a focus on systematic application of risk management techniques. One aspect of risk management is creating a culture of curiosity on how to improve installation operations so that time is collected early in the project, much like contingency budget funding. The premise that unforeseen and uncontrollable events will impact the delivery schedule is the driver of this approach. The most effective means to counter impacts of failure are to forecast potential events which create the greatest risk in project delivery, engage in pre-planning of the risk activities, and track progress against the schedule so that adjustments can be made before time expires. A culture of curiosity is not complete until lessons learned have been discussed with the team and formalized for future reference which can benefit non-team members within the practitioner's organizations.

> **THE "R" Word is Risk Create a culture which asks what is at RISK during all phases of the project**

The perspective of "time" as a commodity must be equally important as budget/cost considerations. In the case of large projects, time can literally be significantly monetized based upon the cost of general conditions by the contractor. One of the projects included in the case studies of this text incurred contractor's general conditions cost of $300,000 per month or $15,000 per

> **Time & Money are the Foundation of Successful Construction**

working day. Mitigation of time can result in savings to the owner at a quick rate based upon the metrics described above, and conversely, increased cost. In this example, these costs only reflect that of the contractor. Other team members, (architects, engineers, and owners) also have personnel that are impacted by delays which result in increased cost. These potential cost overruns are typically internalized against production profits, unless a scope change occurs and there is a condition to charge the owner for additional services. The point is that all team members are fiscally affected by time management and should be included in risk identification and minimization.

Typically, time impacts are beyond the contractors' control, therefore it becomes equally important to communicate schedule risks with the entire team of practitioners (architect, owner, engineer, subcontractors) to facilitate engagement of as many resources as possible during the problem-solving process. Mutual trust between practitioners becomes a "cornerstone" to facilitate this process. Periodic communication, in the form of schedule reporting, serves as the "brick-and-mortar" to build this trust so that transparent discussion can be exercised concerning the project time risks and requirements. Beyond risk forecast, this reporting process also includes successful mitigation efforts which aids

> **TIME TRUST is created by transparent and consistent communication between project team members.**

in building trust between all parties by demonstrating engagement in the project efforts to create success. In the event that the time lost is within the contractor's control, this form of reporting is an effective means of training the team to perform at a higher level, since repetition facilitates learning. This communication report has been implemented by the author for 15 years and will be discussed in depth in further chapters.

The content in the following chapters will provide a framework to better identify schedule risk, strategic concepts on improving time delivery, and how to engage all team members in the process.

Chapter 1

Has It Ever Gone As Planned?

After 38 years of direct and indirect scheduling practice, the author has yet to see a project completed exactly as the originally prescribed plan. One observed common fallacy between all project practitioners is clinging to the premise that they have a perfect plan and further adjustments are not required. This perspective is an illusion that will slow the process in recognizing risk factors and changing circumstances, which ultimately minimizes the team's ability to respond

> **The perfect plan is an illusion, Change is inevitable.**

effectively. A real-life example of this mentality is demonstrated when the schedule becomes "wallpaper." How many times have you walked into a construction trailer, seen the schedule attached to the wall, and peeled the corner of the schedule back to find the wall faded behind it? This condition is almost always an indicator of the practitioners viewing the schedule as static and non-responsive to changing conditions.

1.1. Change Is the Only Constant

As in life, change is inevitable and will occur during the life of a project. Success is measured by how the team adapts to the dynamic environment and the schedule reflects the time changes associated with change response. The organic nature of the schedule, whereby it consistently evolves due to changing conditions, is essential for

> **Project Success is Hinged on the Teams Ability to Adapt**

project practitioners to accept. Change can appear in many forms: staff turnover, owner-driven scope change, local regulation requirements, environmental impacts, and end user's inability to conceptualize two-dimensional drawings. Actual experienced examples of these types of conditions are as follows:

- Staff turnover and owner-driven changes: The week prior to scheduled construction groundbreaking, a key leadership position of the owner's team was changed (Director of Facilities and Construction) which resulted in re-evaluation of the projects program to meet the end user's needs. The design and construction team were placed on hold for three months during this evaluation period. Upon completion of the program review, the decision was made to commence construction immediately and add additional floors to the project. This change in direction impacted the entire team of architects, engineers, and contractor. Actual details of this schedule impact are discussed in the chapter for case study 1.

- Local regulation requirements: During the final week of inspections on a five-story high school, the mechanical inspector determines that a potential fall risk could occur when a heating, ventilation, and sir conditioning (HVAC) unit will be serviced on the roof. With one week left to go in the project, that inspector decides that a final certificate of occupancy will not be granted until permanent handrails are installed. In this particular case, the team had built contingency time into the schedule so there was no impact to the schools opening day.
- Environmental impacts: A school of nursing building, located in the Texas Medical Center on a highly congested intersection, was in the construction process when an optic fiber cable was discovered inside the building line of the foundation. The cabling wasn't identified on the plans nor was it labeled as to whom it belonged to. Since the impact of critical care services could be at risk, by cutting the line, a 2-week delay occurred to the foundation while the project team determined which institution owned the line. As this was early in the project, there were multiple opportunities to mitigate the delay and finish the project on time.
- End user's ability to understand two-dimensional drawings resulting in nonalignment of expectations: This type of nonalignment event probably occurs most often across all types of projects because end users are typically not well-versed in construction application. To illustrate this situation, a recently constructed five-story high school was designed with structural concrete frame and floors. The architect designed the floors to be polished concrete and the final appearance of the floor wasn't communicated clearly enough to the end user. When the building was almost ready to open, the end user didn't like the fact that there were natural cracks in the polished concrete, a condition created by the nature of elevated structural concrete. This late discovery was the result of not clearly aligning expectations of the finishes, which could have been avoided early in the project. As construction practitioners, it is critical that this type of knowledge is not taken for granted, and the end product is visualized through the end user's perspective.

As change is a given, it is equally important to understand that the impact of change is less when incurred earlier in the project. Figure 1 illustrates the impact on cost relative to the effectiveness of change during the lifecycle of a commercial construction project. Most practitioners are familiar with this illustration; however, the aspect of time is seldom addressed.

In reality, time follows the "Effectiveness of Change" path. The "less-more" axis becomes effort. When greater effort is applied earlier in the project (Predesign and Schematic Design), probability increases for the creation of time contingencies and minimizing risk. The primary lesson learned with regard to change is anticipate as much as you can, as early as you can, so you have reserve time to utilize for those unanticipated changes.

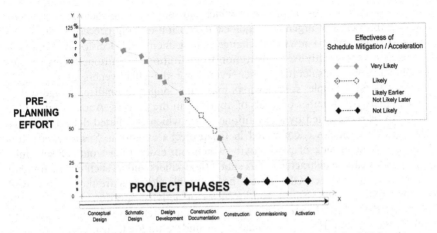

Figure 1: Effectiveness of Change in the Form of Schedule Mitigation or Acceleration during the Life of a Project.

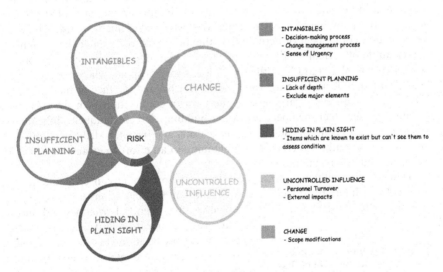

Figure 2: Forms of Project Risk.

1.2. Forms of Risk

While change is considered the most common bearer of RISK, other aspects of project management can impact a schedule such as intangible elements, insufficient planning, uncontrolled project influence, and elements that are hiding in plain sight (Figure 2).

The illustration in Figure 2 above, represents high-risk aspects which can impact a project timetable and share the trait of potential change. Intangibles are nonphysical states or processes which can impact time such as an attitude lacking in a sense of urgency or inefficient decision-making processes. The experienced practitioner knows that change management and decision-making processes can vary greatly from institution to institution. Practitioner firms are not excluded from consideration when evaluating the effectiveness of these processes. As an example, some owners may take longer to finalize change orders due to increased number of levels of approval in the process. Another intangible risk can be represented by a practitioner team which is deflated due to excessive changes or obstacles encountered in the project's execution. Sustained frustration can lead to lack of productivity which can erode time from the schedule. Intangibles can be characterized as subtle time-killers and as such can be difficult to identify and correct. Oftentimes, the intangible elements are those that can be the most destructive to time contingencies or activity durations because of the nonphysical nature. In preparing a schedule, these items are difficult to quantify on a time scale and oftentimes ignored or underestimated. Insufficient planning is a tough risk factor to evaluate by most execution teams since practitioners tend to plan based upon past experience. While the approach of utilizing past experience and lessons learned is considered as a positive, it can be limited in not providing enough depth. A good rule of thumb for the level of depth to apply in the planning effort is to equal the perceived element of risk with the construction component or activity, i.e., the more risk, the greater the planning. An example of matching risk with the level of planning was the construction of a steel crown, aka "tiara," on the top of a 25-story hospital, where the structure was elevated 40 feet above the roof and 25 feet beyond the face of the building. The unusual nonrectilinear shape of the tiara in conjunction with the size and location of the steel members made the building element uniquely difficult to construct. As a result, the designers, fabricators, and installers dedicated approximately 4–5 months of collaborative planning in an effort to ensure an efficient schedule and safe installation.

Uncontrolled influence can appear in many forms and can originate either externally or internally. One of the most common manifestations of uncontrollable impact to construction is catastrophic weather events. Tornadoes and hurricanes can occur with relatively short notice, preventing preparations to protect the construction site from damage, unless of course your project consists of interior scope only.

> **Effort of Planning should be equivalent to the level of Risk**

The key word in identifying this type of risks is "uncontrollable," which can manifest from employee turnover or as contractual language defines the term *Force Majeure*. The Latin derivation of the term is superior force beyond the control of all parties and instances including crime, war, weather, and labor strikes. The last and sneakiest construction risk element are those issues that are hiding in plain sight. These risks are known to exist, but their condition cannot be assessed due to their lack of accessibility. Two conditions of the most common construction examples are in the form of inaccurate "as built" drawings for a remodeling project or incorrect depiction of underground utilities. In both conditions, the practitioners believe they know what exists, but in reality, the final determination cannot be achieved until the existing conditions are uncovered. One of the most common mistakes on construction projects is failure to uncover existing utilities early enough in the schedule to allow for potential correction of existing conditions. One such condition occurred with an existing sanitary sewer line in a congested urban setting, where two independent agencies had modified the lines flow. Upon investigation, it was determined that the last agency ("B") who modified the line had failed to notify the other agency ("A") of the change. The civil design professionals utilized agency A's as-built drawings, which was typical practice. In this instance, the existing sewage system was stagnant with the manholes holding 6′ depth of waste. The contractor had field verified the existing utilities early in the project and there was adequate time to redesign the utility lines to accommodate sufficient flow for the sanitary sewer system. This situation occurs more frequently than expected and has shaped the risk approach of "what you cannot see, will hurt you." The purpose of this phrase is to accentuate the importance of exposing nonvisible conditions earlier in the schedule to allow for time contingencies.

With all the forms of change illustrated above, it could be considered common sense to focus construction planning on elements of risk. Statistically speaking, however, the higher risk elements have the highest probability for change, therefore time spent on contingency plans may yield the best schedule results. The conservation of time, especially in the early stages of a project, is critical to schedule management. In technical terms, this time contingency is referred to as float. In practical terms, it is best to think of float as a time contingency, which can be applied for mitigation of the risk factors in Figure 2. Team success is differentiated by how we manage risk and solve problems. How we respond to change defines us as practitioners. And regardless of the practitioners technical scheduling prowess, everyone relates to risk. Just as it takes multiple perspectives to create a more complete understanding, it is key to realize involving all team members in the quest to identify and assess project risk will contribute to a better execution plan. This collective, risk-focused approach shifts away from the old school paradigm of scheduling as only a technical exercise in illustrating construction activities. The technical aspect of scheduling is still important but the focus on risk allows for expansion of content. Acceptance of this method facilitates understanding high-risk activities from all members of the project team during all phases of the project (i.e., predesign, design, budgeting, procurement, permitting, construction, commissioning, and end user activation) and creates a comprehensive plan. These are the reasons why this book focuses on the art of managing risk in construction.

Responsiveness to changing conditions after the creation of the baseline schedule is a key characteristic of optimized schedules. With all of the potential for change, as identified earlier in this chapter, thought should be given to how the baseline schedule is organized. This approach is often neglected by practitioners during the development of the original schedule due to a myopic focus on content. Those who do consider the organizational impact generally limit their layout to a work breakdown structure (WBS). Specifically, the schedule should be created where activity adaptations can be easily implemented during the life of the project to reflect change. This planning should consider the constructability sequencing of the project. An example of this approach was implemented on a university classroom and administration building which required all exterior walls to be replaced due to leaks. The situation was compounded as the building was in the shape of a "piano" and had approximately 12 different elevations. Other constraints on the building process was the building had to remain operational during construction. Early in the planning, it was apparent that the glazing subcontractor may have worker staffing issues which could result in fluctuations of productivity. Based upon these high-risk levels, each elevation was planned as an independent work sequence. Once the activities have been developed, the schedule was organized in a similar fashion, by elevation. Within two weeks of starting actual construction, the subcontractor confirmed that worker availability had changed and required re-sequencing of the elevations. Due to the anticipation of this potential change, the team was able to adapt the schedule within a short period of time and still met the overall delivery mandate. There were additional changes invoked by the university to accommodate off-cycle occupancy needs. The approach of anticipating change as it reflects to the organizational structure of the schedule facilitated a "win-win" scenario for the entire team.

The last critical aspect of schedule risk management involves understanding the difference between contingency and mitigation planning. Merriam-Webster dictionary (2017) defines a contingency plan as "a plan that can be followed if an original plan is not possible." Mitigation (Merriam-Webster, 2017) is defined as lessening the severity of damage or loss. In the context of time, contingency implies that it is prepared prior to actual impact to the baseline schedule, where mitigation seems to apply an ad hoc approach to minimize delay impacts. In terms of the final outcome, mitigation assesses the results of the contingency plan. This distinction is important as the higher risk implementation processes should have contingency planning prepared prior to actual execution. Mitigation becomes more of an immediate adjustment approach to managing time delays.

In summary, with every project, there is always RISK. The sources of risk can vary from internal to external, forces of nature, labor, economics, politics, project leadership, delivery method, and so on. The constant in project execution is change. Risk planning is necessary once the baseline schedule is prepared to increase the percentage of successful deadline delivery. While many of the concepts described can be correlated to common sense thinking, the point is that practitioners can increase success by adopting a risk assessment approach. In order to apply this paradigm, collaboration and risk assessment should be at the forefront of all planning efforts.

Chapter 2

Who Benefits from Planning and How?

To grasp the extent of benefits from planning, the distinction between planning and scheduling must be understood. There are some key distinctions that differentiate the two efforts. The first of which is that any stakeholder can participate in planning, but scheduling is limited to practitioners who create the schedule utilizing a software platform. Knowledge of the construction process is not required during planning, where it is mandatory during scheduling operations. Planning addressed the question of "Who?," "What?," "Where?," and "How?" while scheduling answers for "When?". Planning is generally framed from a macroscopic perspective vs the microscopic characterization of scheduling. The one common aspect of both planning and scheduling is RISK. These differentiators are presented in a broad perspective. The following paragraphs will delve into the smaller, more specific elements of planning and scheduling.

2.1. Planning

The most significant aspect of planning is that it is inclusive of all stakeholders and their knowledge of construction methodology does not prohibit them from participation. The members of a team can range from owner, designers, general and subcontractors, utility providers, neighborhood associations, end users, maintenance personnel, and regulating authorities. The key parameter is that each person understands or represents some element of risk to the project. The significant questions become "What is a risk to the project?" and "What is the worst case scenario?". To frame your thinking in terms of risk, below are examples from differing team member perspectives (Table 1).

> Planning answers Who, What, Where, and How?

Each of the questions presented in Table 1, are real-life examples of schedule impact and mitigation to preserve on-time schedule delivery. The common element from each stakeholder's perspective in Table 1 is time. And each of these time periods pose risk to the overall schedule. In the owner's case, there was a job which had an extraordinarily short time period for completion. The building was a complex medical facility in a suburban environment. Due to competition in the area, the time frame for completion became fixed. In this case, the contractor was aware of the situation and had to perform a herculean effort to deliver the building on time. In this particular case, costs were higher as expedited delivery and installation became the norm, not the exception. From the designer's perspective, sole source delivery does not create an issue because they

Table 1: Stakeholders' Potential Risk.

Stakeholder	Potential Risk Questions
Owner	• What is the impact to ongoing operations in adjacent facilities?
	• Is there a fixed requirement for project completion?
	• Is project financing dependent upon a fixed completion date? If so, are there penalties for missing the date?
	• Are there easements which must be obtained from adjacent land owners or utility firms?
Designers	• Is a material provider a single source closed specification?
	• Are there elements of the new building that must tie into an existing building? If so, does it involve MEPF, Integrated Technology (IT), or structural connectivity?
General contractor	• Is the staging area adequate for construction? If not, what are the options nearby? How will personnel and material deliveries impact the schedule?
	• Are all of the subcontractors strong performers? If not, which one needs to be shored up to optimize performance?
	• What are the long lead items?
	• Will special equipment be required to execute the project? If so, is it readily available?
	• What is the proximity of the new building to existing structures? Are protection measures required?
	• Will the foundation and structure operations of construction be subject to harsh weather conditions? If so, what can be done to mitigate loss delays?
Subcontractor	• What are the maximum number of personnel required during peak construction periods? What is the production loss for getting the workers to their work space for elevated construction?
Utility provider	• Will new service lines be required for the project? If so, how long will this take and who bears the cost?
	• Does an existing utility line interfere with proposed construction? If so, how long will it take to relocate the service line?
	• Are the existing service's sufficient to supply the new project?

Table 1: (*Continued*)

Stakeholder	Potential Risk Questions
Neighborhood associations	• Will the facility elevation be higher than neighborhood fencing? If so, will there be an issue of privacy violation?
	• Will noise from structural construction reach the adjacent homes? If so, what hours of operation may be impacted.
End user	• Will mock-ups be available of critical function area so the final product is clearly delineated? When can they be available for viewing? How does this impact the schedule?
	• Can an end user shut down construction operations?
Maintenance personnel	• Are the proposed monitoring control systems for the new project compatible with existing system?
	• How long will the training period last for eventual control of the new facility?
	• Does the maintenance dept. have sufficient staffing to handle the new load of maintenance for the new building? If no, how long will that transition period be affected?
Regulating authorities	• Is a permit required for the project? If so, what is the timing to procure a permit?
	• Are there regulating entities specific to the business operations that must provide inspections and approvals? If so, what is the process and how long does it take?
	• Does the owner provide inspectors to verify installed work? If so, what is the notification period?

are looking for a unique product to give the building a specific visual appearance. However, in the contractor's world, sole sourcing typically creates time delays. The lack of competition between vendors can create an attitude that they can deliver the product when it is convenient for the vendor. During a long spanning career, 38 years have shown that delays are typical when singular supply sourcing exists, no matter how good intentions may be. This condition should always be considered a high risk and contingency plans are a must in constructability of the product's surrounding area. The significance of the Stakeholders' Potential Risk table is that every project has its own unique set of risks with similar participants. To maximize contingency planning and early risk mitigation, it is essential to collaborate with each stakeholder and ask the risk questions that relate to your project. Once the questions become identified, then stakeholders and the team practitioners can begin the process of addressing each

**Capture Risk Elements
from Multiple Perspectives**

risk in an effort to minimize impact on the construction schedule. One of the primary purposes of writing this book is this process of reframing risk from multiple perspectives. It is common that practitioners include construction risk sequences in the schedule, but typically only from the perspective of the contractor. This viewpoint is limited. In

**What You Can't See Will
Always Hurt You!**

one of the previous chapters, the concept of "what you can't see will always hurt you!," These stakeholder viewpoints fall within that categorization. If the project planner doesn't understand the issues that accompany stakeholder risk, then the scheduling effort is incomplete and subject to failure.

Acquiring this information is as simple as conducting a sticky note session or interviews. Sticky note sessions are typically the most productive where each stakeholder is given a sticky note pad and asked the question "What is your greatest risks in the execution of the project?" A facilitator then places each concern on a wall board, categorized by the phases of a project (pre-construction, construction, commissioning, activation). Each element of risk is discussed with the entire team and collaborative discussion attempts to create a mitigation

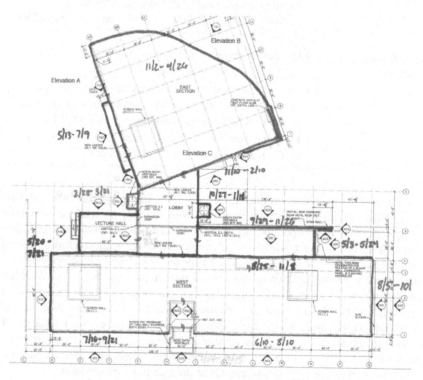

Figure 3:   Multiple Elevations Project (Plan View).

plan. Now, the part which most practitioners fail to compete is translating this information into actual activities in the baseline schedule. In other words, each step to mitigate the condition should be incorporated into the schedule. If this is not done, then the team has effectively ignored valuable risk and mitigation information. An example of this process was on a project where operations were ongoing for 10 months out of a year. The building was housing both administrative and student personnel and was at the core of a mid-size campus (Figure 3).

Construction scope included demolition and replacement of the exterior skin of the building. The exterior walls consisted of 14 elevation surfaces with 2 entries in and out of the building. There were windows of opportunity where the building would not be occupied, primarily between semesters or during midterm breaks. With so many variables in the work conditions and the risk of exposure to interior spaces during construction, the planning phase became critically important. The primary questions were "Where is the starting point?" and "How can flexibility be integrated into the schedule?" As with any planning effort, the best approach is to break the scope down into the smallest components. This drill down process typically frames the sequences into more manageable units by practitioners. In this example (Figure 4), the units are represented as wall elevations. In this particular case, each elevation was identified on a large sticky pad in a conference room with a photograph of the existing elevation at the top. The team then identified the risks associated with each elevation by adding them as bullet points on the sticky pad. Once this was complete, the activities were developed for each elevation. Logic between the activities was the next step.

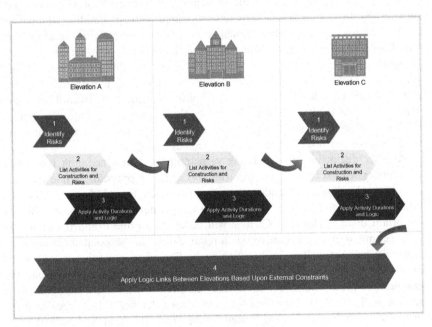

Figure 4:   Risk Process.

Each elevation was treated as a separate grouping of activities within the schedule. At this point in the process, the constraint variables imposed by the needs of the building occupant (owner) can be applied. It is important to understand that grouping the activities in this way allows flexibility for future changes. For this situation, these variables would be timing restrictions based upon user occupancy. The planning effort described above included participants from ownership, general contractor, subcontractor, program manager, and end users. One interesting note for this project was two weeks after agreement on the baseline schedule, the subcontractor advised the team that their resources were insufficient to complete as planned. Because the schedule was planned in the method described above, the changes required to modify the schedule were easily implemented.

2.2. Scheduling

Scheduling can be characterized as the "when" aspect of the project plan and is considered the technical application of planning efforts. With the progression from planning to scheduling, several elements move from macroscopic to microscopic in characterization (Figure 5). Technical knowledge and level of detail move from less to more with the progression process. Range of participants and perspectives shift from wider to narrower as planning shifts to scheduling. The level of organization also becomes more refined during the scheduling process with the implementation of work breakdown structure (WBS). For example, all activities associated with the building foundation are typically grouped together. Similar categorization of activities is assembled based upon structure, exterior skin, mechanical, electrical, plumbing, et al. With the specific application of time in the form of activity durations and relationship logic, the schedule reveals the overall duration. The critical path identifies the activities that cannot slip in duration without impacting the overall schedule. This narrowing focus on a specific set of critical activities enables the practitioners to focus on a smaller group of steps in order to stay on schedule. As issues become narrower, manageability becomes easier.

Scheduling answers When?

Another distinction of scheduling vs planning is that scheduling requires a practitioner who is capable of operating a scheduling program. While the purpose of this book is not to endorse any singular scheduling platform, it is important to understand that a good software operates as a database. This is important due to "change." Change will occur on a project, and the ability to manage time is greatly enhanced when relative data can be isolated for analysis. Database protocols allow information to be filtered so practitioners can focus on the issues. An illustration of this concept is a schedule for a precast parking garage. The precast supplier was initially directed to deliver the materials by a specific wall elevation. Unfortunately, the precast detailer did not receive those instructions, so the panels are delivered for the wrong elevation. Suddenly, the precast installer must change the erection plan. The project team only needs to view

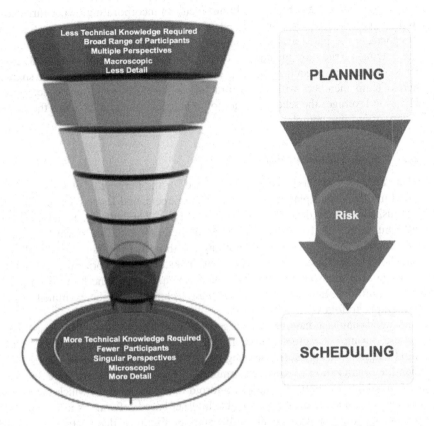

Figure 5: Progression Comparison From Planning to Scheduling.

the schedule that was originally created for precast. A database program will allow this data mining technique, so practitioners can resolve the situation quickly.

The one common element between planning and scheduling is risk. A typical mistake made by practitioners is failure to translate the risks discovered in planning to the schedule. This condition is often created due to exclusion of activities whose ownership originates by other team members. Another way to say it, is the contractor who generates the schedule is only interested in what they are responsible for. In reality, the contractor is contractually bound to delivering a project on a specified date therefore, it is advisable to include activities generated from other team members. This is important because delays generated from sources beyond the control of the contractor are allowable per typical contracts. In essence, the contractor is the manager of the complete delivery process, even if execution is not their primary responsibility. This is a fundamental concept to all construction delivery methods which is why it's

Risk inclusion leverages the schedule as a tool to manage time

perplexing that more contractors do not focus on incorporating owner-furnished and owner-installed elements in the schedule. The same holds true for the designers, if the plans are incomplete at the time of project commencement. If these items, or risks are included in the schedule, then it becomes inherently easier to manage the execution process. This method enables risk to become shared across team members with the contractor as the execution team leader. Risk inclusion leverages the schedule as a tool to manage time and increases the probability of on-time completion.

2.3.   Risk Identification

There are many methods that can be utilized for the purpose of identifying risk. Mind mapping, pull planning sessions, white papers, and lessons learned are all established tools to develop project specific risks. The author has developed a risk mitigation matrix for the purpose of determining risks during the pre-construction period. The objective of any collaborative tool is to engage all the stakeholders and systematically illuminate risks and assign priorities. An additional goal is to examine worst-case scenarios and determine how to mitigate these conditions. Examples of each of these techniques will be examined in the paragraphs to follow.

A mind map is a diagrammatic organization of information and is best-suited for use during conceptual planning. Due to the nonlinear aspect of this tool, mind mapping facilitates free flowing ideas. Once these ideas are on the canvas, then the organization process can commence. There are many software applications available for both iPhone and android operating systems which facilitates convenience in short bursts of thought. Internet-based software is also available for implementation in a collaborative setting. Figure 6 demonstrates a mind

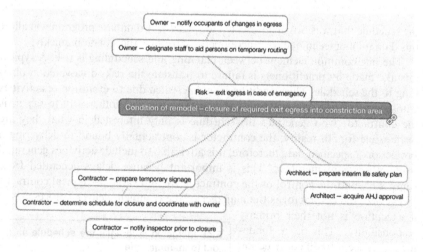

Figure 6:   Mind Map.

map for a specific risk condition of providing exit egress in an existing building during renovation. Actions to address the condition of risk emanate from the situation (illustrated in the center of the map). Notice that colors can be used to easily differentiate responsible parties. This technique is typically quick to perform. Since this is a nonlinear method, the drawback is that timing aspects have to be applied through a different means. Additionally, mind mapping is generally limited to one concept per map. In summary, this tool is characterized as follows:

- Participants: both singular or collaborative,
- Ease of use: very intuitive.
- Tool availability: takes a little effort to locate a preferred electronic platform.
- Best use: planning.

Pull planning (also referred to as sticky note sessions) is demonstrated in Figure 7. This particular example shows the sticky notes in the background with the foreground representing the results in a flow chart. This method is particularly useful in a collaborative setting with multiple contributors providing different services. In this example, the purpose was to identify the activation and commissioning process for a complex heating and cooling plant. The different color notes represented a singular contributor (i.e., light blue represented the electrician, white represented the mechanical contractor,

> **Pull Planning Breaks Down Knowledge Silos Between Project Participants**

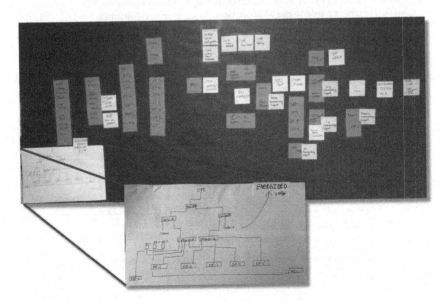

Figure 7: Pull Planning (aka. Sticky Note Session).

and turquoise originated from the plumber). The exercise starts with each firm independently writing activities on a sticky note. A facilitator then leads the process by asking for the sticky notes in a chronological order. During the placement of each note, the facilitator discusses the content with the group. This process is particularly effective in sparking additional ideas from the contributors and sharing information across dissimilar trades. This process is then repeated for each member of the team. The addition of each note provides development of logic and creates a better understanding by all parties of the full scope of work to be performed. The results can then be translated into a schedule with durations for each activity and reflective of the logic identified during the exercise. In summary, this tool is characterized as follows:

- Participants: collaborative.
- Ease of use: a little more thought is required for this tool as both step identification and linear sequencing are involved.
- Tool availability: readily available with wall paper and sticky notes.
- Best use: planning and scheduling of complex processes or when multiple resources will influence the outcome of the scope.

1-Line Pier Conflicts

Construction Co. has been asked to evaluate the cost and schedule impacts associated with three separate options for the conflict that exists along Grid Line 1 between Grids D and G. The conflict was discovered while excavating the top seven feet of material in this area and large piers were discovered on top of Grid Line 1.

Below is a summarization of three options.

Option 1: Install bigger and deeper between the old piers

A. Summary: This option requires that bigger, deeper piers be drilled between the existing piers. A deeper cap beam will be required to bridge over the old piers. This option allows the basement wall to remain in one plane (no bump-out).

B. Pros and Cons

Pros	Cons
Allows the basement wall to remain in one plane (no wall bump-out)	Most expensive option.
	Schedule is impacted due to the need for specialized drilling equipment.
	The sequence of drilling and excavation would have to be revised to accommodate this option.

C. Range of Magnitude: $600,000 to $650,000
D. Design Fee: (To be determined by others)
E. Schedule Impact: 4 Weeks
F. Recommendation: Rejected: This is the least preferable option due to schedule delays and costs.

Figure 8:   White Paper Report.

White papers are ideally suited for presenting multiple options to an aspect of project risk. The context of Figure 8 was an unforeseen obstruction to the foundation of a project. The figure only shows one option (refer to the appendix for the full white paper with three options). Essentially, the risk is portrayed in pros and cons with associated cost and time impacts. Additionally, the author of the white paper also includes their recommendation on the option under consideration. This is a particularly effective tool for use with nontechnical decision-makers as the risk is described in common terminology. In summary, this tool is characterized as follows:

- Participants: singular.
- Ease of use: most complex of the tools described as both time and cost considerations must be identified.
- Tool availability: readily available.
- Best use: planning and scheduling when evaluating multiple options.

Lessons-learned assessments (Figure 9) are typically implemented in learning cultures to educate practitioners on areas for future improvement. This memorialization of challenges, recommendations, and successes can serve as a basis for identifying risk on similar projects. For instance, one of the challenges was related to room numbering for the building. In this particular case, the numbering of the rooms was performed late in the project. The lesson learned was that it impacted the labeling of all the MEPF (mechanical, electrical, plumbing, fire sprinkler) systems, since the operator of the building required each system be labeled to the corresponding rooms they served. The later delivery of this item in the project caused some timing delays that had to be mitigated during installation and commissioning of the project. The benefit of this deliverable is that reference to the document can raise awareness as to potential risk, especially to practitioners that were not on the project team. In summary, this tool is characterized as follows:

- Participants: collaborative.
- Ease of use: simplistic in nature but requires a facilitator to direct conversation.
- Tool availability: readily available.
- Best use: post-project completion to assess risk points during the project and derive methods to avoid in future projects.

2.4. Benefits

Each of the planning tools described in this chapter take time to implement. In order to justify the time, most practitioners want to understand the advantageous aspects of these processes. One method to assess the gains achieved through project planning, is to understand the negative results when planning is not utilized. Time slippage is more often than not the result of inadequate

LESSONS LEARNED SESSION NOTES

CHALLENGES

- Commissioning-spend more time with the commissioning scopes, based on what would be commissioned and what the MEP's would include.
- There was no clear definition of what the scope would be.
- Restack/Room numbering.
- Permits – Have a better set of drawings to give to the City of Houston.
- Mockups
- Actual Equipment Budget–Medical/IT
- Owner to identify who was to move to OPC - Continuity/Communication.
- IT/NC/Security under Architect.
- Have operational sooner. IT backbone installed sooner.
- Changes in technology – IT/Medical Equipment.
- A & E better analyze Value Engineering before accepting.
- Dirt Doors in Infusion.
- Medical Equipment Purchasing.
- OR/HVAC requirements to Ortho (Timing)
- Elevators
- Window Washing Equipment
- Vibration/Isolation on the transformers.
- Central Plant on the 17th floor (New 20)

RECOMMENDATIONS

- Identifying commission scope early.
- Renumber floors/room @restack. Wait to label MEP until after renumbering.
- Procure foundation permit. (Permit the shell so that construction can proceed). Then procure core and build-out permit.
- Prerelease MEP Equipment.
- Spend more time with mockups – to implement to construction. During design and development.
- Owner review of Medical Equipment to be moved or revised to help identify budget. Have a better idea of what is available for reuse/disposal.
- Owner to meet with end users to review function of new spaces, not just directors.
  - o   Details
  - o   Data Power
  - o   Hardware
  - o   New codes (Requirement)
- Designated Advocate from each department.
- Plan reading (Mini Maggie Team)
- Architect to have IT/NC/Security
- Medical Equipment/IT on same plan.
- IT Backbone/IDF Room/Biomed coordination sooner
- Coordinate "TMH Standard"/Code compliance requirements.
- EHS Involvement sooner.

WINS

- Mock up on waterfall
- OR conversion to Orthopedic
- Exterior Lighting
- On time/Within budget
- Overwhelming Patient Satisfaction
- Bistro (Very Profitable for TMH)
- Truck Rodeo
- Hurricane Plan (Drills to prepare)
- Equipment delivery coordination (Move in)
- Furniture Selection – Yeah Lori!
- Art
- Mitigation of Medical Equipment Changes
- BIM/3D coordination process
- Coordination with selected vendors
  - o   I.E. Curtain wall
  - o   Elevators
  - o   MEI

Figure 9:   Lessons Learned Report.

planning. Project stress becomes the subsequent manifestation of delays that impact the critical path. Prolonged periods of stress almost always undermine team confidence and motivation. Collaboration can become diminished as team members may blame one of the practitioners for the delay, sometimes unfairly. A noncooperative attitude can extend beyond the management problem-solvers into the field where worker coordination across multiple trades lessens. Field crew agitation may result in less communication thereby impacting coordination of multiple trade installations. Mitigation means of extensive overtime or multiple shift work can decrease productivity in production. Safety may also be negatively impacted with an increase in infractions due to worker tiredness. Reduced morale can lead to less ownership of craftsmanship, thereby minimizing quality. Project cleanliness can also be impacted which creates less perceptible delays in material management in the worker's environment. Each of these manifestations of stress chip away small increments of time in the production cycle. These "microdelays" are less identifiable as they are not necessarily attributed by an activity within the project schedule. In some ways, these types of delays are more dangerous to the schedule because of their elusive nature.

Conversely, proactive planning has the potential to improve productivity, barring unforeseen conditions. Productivity can result in higher profit margins based upon shorter overall durations and less cost. Construction projects with minor or no delays create a conducive environment for architects, engineers, program managers, and contractors to meet original budget forecasts. Staff stress exists at a minimal level when projects can function at this level. Less anxiety cultivates creativity in problem-solving and a positive perspective when issues arise. These factors increase the probability of a profitable project by all practitioners. The ultimate goal of the owner occupying the building early or on time thereby meeting or exceeding revenue projections, creates a win-win for all.

Another benefit from planning is that engagement in the project increases by those team members who actively participate. Multiple perspectives and familiarity with the project increase the percentages of effective planning when performed in advance of actual work. This engagement also facilitates buy-in from the practitioners which solidifies commitment to the execution plan.

Thirty-eight years of experience has born witness to each of the conditions described above. Project stress cannot be overstated as a time deterrent. One of the more notable examples of project stress impact was occurred when unknown underground utilities had been discovered all around the site after the foundations had been completed. These utilities were high voltage in nature and were located in the pathway of new underground electrical services. Due to the sensitive nature of the existing utilities, the project was delayed for two-and-a-half months while the land area was hydrotested to determine exact locations. Additionally, there was a change in executive management by the contractor during this period. Momentum of the project was stalled early in the project and general condition monies spent with no productivity. While time was extended to cover this delay, the full general conditions cost was not recovered. The architect and engineers still had to pay personnel even though construction administration had temporarily halted. The financial impact as well as lack of activity greatly impacted the morale of the project team. Collaboration between the civil engineer and the contractor became more difficult with each passing day of hydrotesting. A significant amount of energy had to be implemented towards regaining a positive relationship between the practitioners. This situation exposes that relationship issues can erode a schedule and redirect energies that could be used to efficiently solve problems.

Another negative impact example occurred on a different project where an existing building was demolished to allow new buildings to be built on the site. During the design planning phase, the architect had requested all as-built drawings of the existing building. The owner indicated that all existing plans had been destroyed since the original building was built approximately 85 years prior. Upon commencement of foundation construction, the contractor discovered existing foundations below grade, within the parameters of the new buildings. Shortly thereafter, it was discovered that an existing foundation plan actually existed, and it matched what was being discovered underground. While this was not available for planning prior to commencement, the project team used the plan to evaluate other uncovered locations and modify the new foundation design. Due to the large site area, this created a three-month delay which was ultimately mitigated by two months.

This next example illustrates the positive impact of pre-construction planning. The site was located at a major intersection in an urban area. Utilities which served multiple buildings in the area wrapped around this project in the city right of way. The contractor assumed the perspective that due to the density in the area, that existing conditions may not match the as-built drawings. Prior to construction of the plaza area, located in the right of way, the contractor executed exploratory hand excavation to identify all existing utilities. The effort proved to be effective in that multiple fiber optic service lines were discovered as well as frayed telephone banks. Because this effort had commenced well before construction commencement, the structural engineer had the time to redesign the foundation of the plaza utilizing strap beams so that existing utilities would not be touched during construction. The contractor also had time to contact the utility providers so that they could update their as built plans and watch the foundation construction. This created confidence by the utility providers that utilities were not impacted by construction work and provided transference of valuable information. The actual redesign of the plaza foundation produced a savings of approximately three quarters of a million dollars when compared to the other option of relocating the utilities. This win-win situation created a positive environment for the remainder of utility coordination between all practitioners on the project.

In summary, most practitioners relate to the impacts created by actual construction activities better than the intangible processes. In reality, these intangible efforts are just as capable of eroding time away from the project schedule. For this reason, it is important to apply pre-planning and reap the benefits created from the actions. This saved time will serve as a buffer when mitigation cannot be applied successfully. The next chapter will describe real projects and the mitigation efforts that were applied to meet the required completion date.

Chapter 3

Real Construction Mitigation Case Studies

As referenced in the introduction, many of these scheduling concepts have been born from experience through actual project execution. Therefore, it seems appropriate to include a few case studies that illuminate the approach of evaluating and resolving risk during construction execution. In order to provide a full perspective of the project, the risk situation, and the means employed to mitigate the project successfully, senior leadership from the architectural firm, owner, and contractor were interviewed. For the purpose of clarity, the author's role in each project was as the program manager or scheduling consultant. Each case study will describe the project scope, practitioner team characteristics, risks, mitigation efforts, and the final outcome. Of interest, all projects selected for the case studies are large commercial buildings located in Texas cities. The primary benefit to the reader will be to focus on understanding "when is it too late to mitigate?" Secondarily, valuable approaches toward risk management will be presented and may be employed by the readers on future projects.

Case Study #1: Leadership Consistency, Scope Re-Design During Construction, Inflation, And Hurricanes, "Oh, My!"

3.1. Project Scope

The project is a medical outpatient facility located within the Texas Medical Center, in Houston Texas and consists of 26 floors totaling 1.6 million square feet. The scope of the project was new construction on a site with a zero-lot line in an urban setting. The cost of construction was $275 million with a commencement date in 2006. Project before and after photos are shown in Figures 10 and 11, respectively.

3.2. Key Project Milestone Dates

- Architect hired by institution: August 2004
- Program manager hired by institution: April 2005
- Contractor hired by institution: September 2005
- Construction hold by owner   May 2006
- Original construction commencement date: June 2006
  Forecasted completion date was June 2009

- Building program reevaluation completion date: August 2006

  Ownership made the decisions to add mechanical mezzanine, add one floor of surgery, relocation of Heart Center, Central Sterile, and Pharmacy

  New forecasted completion date was November 2009

- Actual construction commencement date: September 2006
- Project scope increase: November 2006

  Addition of three shell floors and roof mechanical mezzanine

  New forecasted completion date was March 29, 2010

- Time extensions granted:

  Hurricane site preparations: new completion April   August 2008
  2, 2010

  Revision for design changes to structural rebar: new completion May 23, 2010

- Owner activation period:

  Move-in commencement personnel   June 2010

  Move-in completion personnel   July 2010

  Actual first patient (projected date August 30, 2010)   July 7, 2010

Figure 10:   Project "Before" Photo.

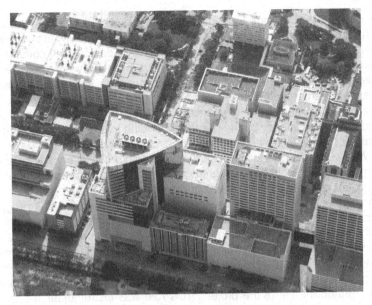

Figure 11: Project "After" Photo.

3.3. Practitioner Team Members

The owner provides health care services to the Houston Region and was established in 1924 with an annual number of outpatient visits more than 800,000 as of 2016 (Houston Methodist Hospital, 2017). The general contractor performs work on an international scale with $4 billion in annual revenues and has been in existence for 79 years (Dodge Data Analytics, 2014). The architect, established for 37 years, also provides design services on an international level and as of late 2014, produced $41 million in billings (Pulsinelli, 2014). Relative to timing, it is important to understand that the architect worked on the project 1 year prior to the addition of the program manager and the contractor. Both program manager and contractor were included in planning efforts at the commencement of demolition and make ready project initiation, which was approximately 1 year prior to the new building commencement date. Early procurement of key subcontractors (mechanical, electrical, and plumbing) was facilitated approximately six months prior to construction commencement.

3.4. Situation and Risks

The first challenge on the project was the owner's decision to re-evaluate the project scope just prior to scheduled groundbreaking. The review of the building program was triggered by changing market conditions and coincided with a change in leadership by the owner. As is often the case with large capital projects, personnel transitions typically occur due to extended duration. Duration for project planning, design, and construction spanned 1, 1, and 4 years, respectively. Probability forecasts would indicate that change is inevitable. Change can be expansive in nature, including many resources and variables required for project execution. In this particular case, the elements of change centered on maintaining leadership consistency by the client, responding to technological advances in medical equipment, and subsequent evolution of patient care.

With respect to personnel changes leadership consistency is critical to all teams (architect, owner, contractor). In this particular case, the vice president of facilities and planning arrived to the management team one month after the project commenced reevaluation of the overall program. This key decision-making position had a significant impact on the direction of the project scope and timing of execution. While change occurred at the start of the project, this role was stable during the evolution of the project. The vice president clearly and consistently delivered the project objectives from re-design through execution. The end user project leader changed once during the life of the project. This role was instrumental in ensuring that patient services met hospital goals and market conditions. The owner demonstrated leadership consistency in this instance by filling this position with a proven member of hospital administration. Additionally, there was a change at the directorship of facility management shortly after project commencement. As was the case with the end user group, the new facility management leader had experience at one of the satellite campuses and was well-versed in the owner's management models. In all three cases of personnel changes, they remained as key collaborators and leaders through remaining life of the project. Leadership consistency consists of providing stable direction so that project goals are easily identifiable to all team members. The best-case scenario of leadership consistency doesn't eliminate change; it improves the percentages for success.

Change also surfaced in the context of restructuring affiliated providers for patient services just a month prior to scheduled construction groundbreaking. This change required reevaluation of program spaces since there were shared services originating from the other providers (refer to Figure 12). As a result, executive leadership questioned the appropriateness of the design program to meet the needs of the institution. Ownership made the decision to postpone groundbreaking three months while the design team and external agencies reviewed the existing program. The project was scheduled to break ground in June 2006, which is the same time that the hold was initiated to commence program reevaluation. In September of the same year, the program evaluation was completed with the recommendation to shift several departmental uses on different floors, add new services, as well as adding a mechanical mezzanine floor to the building. November of the same year, three

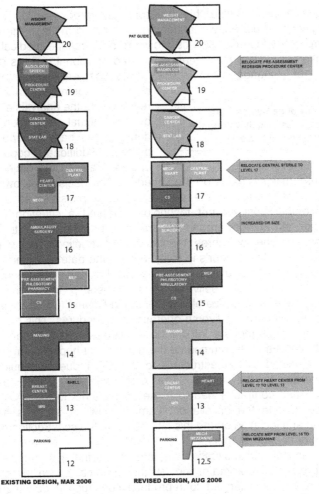

Figure 12:   Restacking Scope Diagram.

additional shell floors were added to the project scope for future expansion. A contingency plan to implement the additional three floors had been incorporated into the vertical structural design of the building from the inception. However, upon execution of the design addition, it was discovered that the structural slab rebar would require adjustments. After the review, the decision was made to simultaneously document design changes as construction proceeded which intensified the coordination between the engineer of record and the

> **The Projects Ultimate Goal is the Clients Business Operations – They Can Create Risk**

construction crew. While this direction generated risk for the project delivery team, it did enable leadership the time to manage the decanting effort from the departing service providers and reshape medical services. Evaluation of system maintenance incurred additional reviews due to the timing of change in facilities management personnel. This diverted some of the team's resources to the analysis effort in lieu of the actual execution process during the early scope change process. Additional scope changes occurred later in the project timeline which had major impacts on the schedule. The decision by ownership to order the MRI units at the very latest date possible, so they could maximize the most current technology created a situation where the contractor had to work around the area and prepare a plan to install the equipment in the last three months of construction. This was further exacerbated by the vendor's inability to meet the date specified as well as the elevated location of the MRI units on the 18th floor (Figure 13). In addition to the MRI's delivery issue, ownership made the strategic operating decision to utilize all the Operating Rooms (OR's) for orthopedic surgery, in lieu of more diverse operational procedures. This decision was made with three months left in the construction timeline and required several changes to accommodate the needs of orthopedic surgery. These needs impacted the infrastructure to the OR's due to requiring lower temperatures in the rooms. The commissioning agent, mechanical engineer, and contractor had to make adjustments to the HVAC system while completing the installation. Laminar flow air systems had to be designed and installed during this late period in the construction process to facilitate orthopedic surgery.

> **Scope Change Systems are Essential, Especially During the Last Phase of a Project**

The secondary risk situation was created by the "just in time" delivery method due to reprogramming of the building space. Design team members were challenged to reproduce construction documents faster than typical durations. The addition in scope of three floors required that the foundation design had to be re-evaluated as well as the wind loads due to the increased building height. Additionally, the restacking of floors created an added mechanical mezzanine which would impact the completed mechanical, electrical, and plumbing service systems design. Since the design reevaluation commenced at the same time of construction start, the architect's projection of six additional months to redesign these building elements became critical. Lapse in the production schedule would create rework of actual construction or delays to the overall schedule, since all of the work was on the critical path at this point of the project.

Figure 13:   MRI Lift to 14th Floor.

The third major risk was fiscal in nature as the inflation rate in Houston, Texas, was at the highest rate in 10 years, at 6%. With the overall duration of the construction phase of the project in excess of 40 months, speed and advance procurement efforts were essential to minimize cost escalation. Limited resources of construction subcontractors, the result of demand in the local construction market, constituted an added element of risk. Mitigation of both conditions, placed emphasis on early procurement in spite of delayed design completion. This excessive demand created further cost increases due to limited manpower unavailability.

Lastly, the environment tested the schedule by incurring delays associated with an active hurricane season on the gulf coast of Houston, Texas. In 2007, Houston narrowly avoided the impacts of Hurricane Dean but in 2008, Hurricane Ike came directly through the project location in the Texas Medical Center. While Hurricane Dean did not cause property damage, there was a delay of four days to the jobsite as the structure was ongoing and had to be prepared for potential impact. Hurricane Dean preparation served as a drill for Hurricane Ike in 2008, where property damage did occur to the project (Figure 14). At the time of Hurricane Ike, the exterior skin of the building was complete through the 5th floor, and mechanical rough-in and drywall systems were in various percentages of completion through the 13th floor. Damage to work in place at the elevated

Figure 14:   Project Damage From Hurricane Ike — 13th Floor.

floors included metal studs ripped from overhead track, sheetrock imploded in multiple pieces due to the wind force, and ductwork sections separated from each other. Water filled all depressed slab areas and had to be removed through pumping operations. Relative to the schedule, the contractor did not request additional time to repair the damage to the affected construction. In retrospect, this recovery effort and the preparedness drill provided schedule mitigation for the repair operations.

The common thread of these risks was their origination point relative to the project team. All issues were beyond the control of the execution team, and yet they were challenged to provide strategic mitigation plans to counter the effect of these impacts. The challenges described above can be summarized in four categories: leadership consistency, scope change implications, escalation cost management, and environmental impacts. The specific methods employed by the project team are chronicled below and offer specific ideas for time mitigation and contingency management.

3.5. Mitigation Efforts

The mitigation efforts have been categorized by the associated risks in order to provide easy reference by the reader. As defined in the previous section, the major risk categories are leadership, scope changes and time management, budget control, and environmental impacts. The mitigation measures are discussed in the following paragraphs and summarized (Table 2).

Table 2: Project Risk and Mitigation Methods.

Risk	Mitigation Technique
Leadership consistency	• Owner engagement • Initiate measures at beginning of the project • Responsive decision-making
Scope change & time management	• Create culture of change management • Establish multiple teams to address changes • Create learning culture through productivity analysis with adjustments to construction process during execution • Create contingency planning at through constructability analysis
Budget control	• Pre-procurement planning
Environmental impacts	• Perform preparedness drills

3.5.1. Leadership Consistency

The initial process that was initiated to address leadership change was the formation of pre-core and core teams to manage risk where the pre-core was a smaller team of highest leadership from the program manager, architect, and contractor. This team kept the big risk items in perspective, forecasted future problems before occurrence, and kept all three team members in alignment with the ultimate project goal, finish on time and within budget. Core team was focused on bringing risk to the owner at the highest decision-making level so that timely direction could be implemented to keep the project within the time and budget goals. Both sets of meetings were conducted on a weekly basis throughout the project. From the architect's perspective, the pre-core team strategy was very effective. Leadership was synchronized with singular perspective for the rest of the team which also facilitated prompt decision-making and strategy adjustments. The owner's perspective believed that early procurement of team leadership greatly assisted the multiple project

> Leadership cohesiveness is essential in creating a low risk project.

adjustments. One year prior to groundbreaking the contractor and program manager joined the team with major subcontractors to analyze and mitigate inflation through early procurement. As a point of reference, the architect was on the team 2 years prior to arrival of contractor and program manager. The owner and contractor reported that commencement of owner activation planning a year prior to actual move-in yielded significant benefits to early use of the building. This planning effort was proactive and involved both the contractor and the owner's end users. Since the building permits were separated by floor (Figure 17), this enabled activation to commence at the completion of each floor, in lieu of the traditional method of at the completion of all floors. The illustration below identifies the actual activation activities by floor as well as medical equipment move-ins (Figure 15). This strategy also created time contingency by allowing the more time intensive floor activities to begin earlier than those floors where activation durations were not as lengthy. Departmental floor designations are shown in the section drawing below (Figure 16).

3.5.2. Scope Changes & Schedule Mitigation

With regard to scope changes and time management, the design consultants were evaluated on their ability to accelerate the design

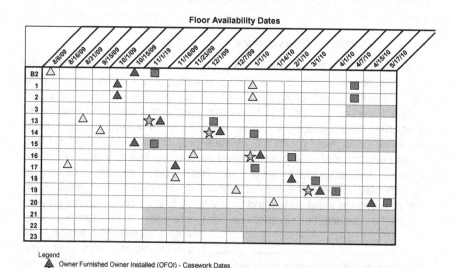

Figure 15:   Floor Availability Chart for Activation Commencement.

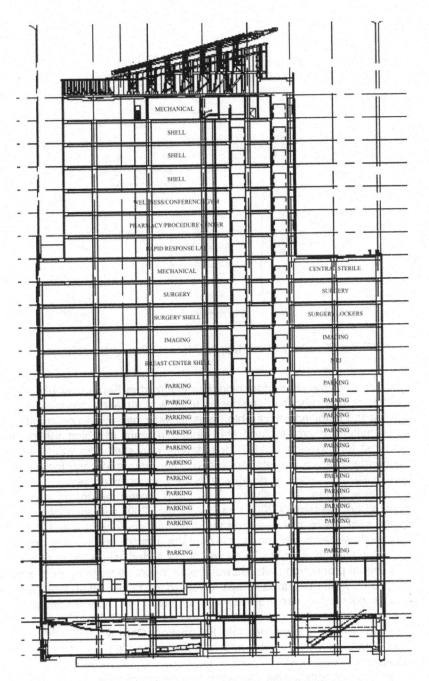

Figure 16:   Floor by Department (Section View).

OPC Permit / Testing Flow Chart

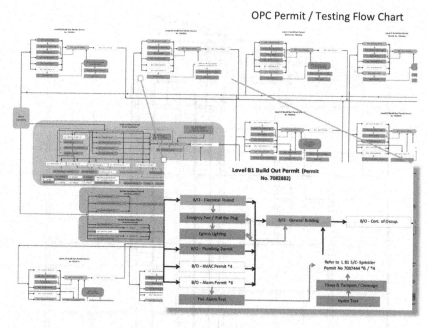

Figure 17:   Permit Flow Map by Floor for Project Completion.

schedule after the reprogramming was finalized in September of 2006. As a result, one subconsulting design team was replaced due to inability to meet the modified deliverable dates to keep pace with construction. Additionally, the architectural staff added a team (doubled the team size) just for redesign of MEP, foundation, and shell who would work independently from the interiors group. This team spent the increased time in the field working directly with the contractor on revisions in a just-in-time model of delivery. This effort lasted approximately six months from September 2006 through January 2007 with the new scope design package by the interior team submittal occurring in March 2008. The multiple team approach allowed the restacking impacts to be evaluated simultaneously with design of the added scope of the project. This split of production units allowed potential problems to be dealt with in a proactive manner to minimize rework in the field. The architect also utilized contractor constructability information on how to integrate changes during the construction process. An additional effort of breaking permitting packages into Foundation, Core & Shell, and Build Out by floor created the time necessary to complete the design due to reprogramming.

While the architect was mitigating the design change, the contractor experienced unforeseen issues during foundation construction as previous foundations were unearthed (Figure 18). This

**Scope Changes Can
Increase Risk**

created a delay of three months which was mitigated by facilitating a faster duration for the retaining wall of the foundation. By changing the orientation of the rebar in the soldier piles, the need for tiebacks was eliminated. The savings in time for tieback construction was equivalent to the duration for excavation efforts of the preexisting foundations which had to be removed.

Upon completion of the design changes, it was determined that the steel reinforcement in the structure required modification to meet the increased floors for the building. The contractor mitigated the delay for modified rebar design for the concrete structure from 12 to 6 weeks. This was facilitated by changing the direction of the formwork for floors 18–25 from clockwise to counterclockwise. This shift in operations facilitated formwork removal at a faster pace; thereby increasing the speed of shell space creation. As the structure was on the critical path, this created a direct reduction in the overall time period of the project. It is important to understand that the increase in rebar quantity and sizing was not foreseen by the designers at the time of the decision to expand the scope by an additional three floors.

Figure 18: Unforeseen Foundation Beneath Future Building.

Production efficiency was created by the contractor by identifying walls in three priority levels: priority walls, secondary priority walls, and all remaining walls. Priority walls were located where the density of the MEP overhead or wall rough in were so dense that the time to construct would be longer than typical wall systems. In order to avoid these walls to become part of the critical path of the schedule, these walls were constructed first to allow the MEP subcontractors more time to construct. Secondary walls were identified as those walls which could be constructed from the inside of the room in lieu of the main corridors. The avoidance of the corridor allowed the overhead MEP systems to be installed in the area first, since this process was longer than the time to construct the wall system. This process created a time contingency which allowed for schedule variances without affecting the overall project duration. Additional production efficiency measures included stocking materials on each floor and clean-up operations during evening, nonpeak work hours thereby enabling the work environment to be ready at the start of each day. Precast erection operations were also performed during evening hours which allowed the crane to be utilized for the concrete structure during daylight hours. A cost analysis confirmed that the continuous operation of the crane was minimal when compared to the monthly savings in general conditions for both the crane and the time reduction created by overlapping activities. Further efficiency was created by sloping the roof deck on the structure so that it would drain and not interfere with construction on the floors below. This tactic was directed at maintaining a work space so that productivity could flourish. Similarly, the ability to get the workers to their respective floors in an efficient time period minimized nonproductive work periods. This strategy was facilitated through early installation of one bank of elevators as well as utilization of two peripheral hoist ways. A designated area for lunch on each floor as well as portable toilets were utilized to maximize productive work periods. The contractor also utilized predictive analysis to monitor and manage productivity to ensure construction execution within the scheduled time frames. Time lapse photography on the concrete framework operations provided the means to analyze productivity and methods so that improvement was applied for the following structural bays. Trending reports for precast panels, glass, and decking were updated daily and reviewed with subcontractors on the same interval to enable adjustments in execution. These real-time feedback processes provided the input necessary to manage construction durations within their allocated ranges thereby keeping the project on schedule. It's also important to

| Production Efficiency Can Create Float |

| Trend Analysis = Real Time Adjustments |

understand that each of these activities were on the critical path of the schedule so any delays incurred would extend the overall schedule on a day-for-day basis.

Prefabrication of ductwork and piping assemblies were assembled offsite and then dropped in place to facilitate shorter installation durations. The entire curtain wall system was prefabricated in panel sections which enabled the exterior skin to finish shortly after the structure was complete. The common denominator in each of the productivity enhancements and mitigation effort was the pre-planning generated in constructability reviews prior to actual construction of each main system. Constructability reviews were conducted on the foundation system, the underground electrical duct bank located in the mat foundation, window washing system, curtain wall exterior skin, tiara structural fenestration on the roof, and truck dock construction which included a mock "truck rodeo." This activity laid out the dimensions of the turning radius's and actually demonstrated operations through real-life simulations from vendors which would service the facility. Each of these efforts flushed out issues that were time deterrents in the building process well in advance of actual construction.

The contractor also utilized the principle of multiple crews to perform specific completion functions which facilitated time efficiencies. Separate and specific crews were developed to facilitate a mock Texas Dept. of Health inspection at 80% completion, so that adjustments could be made prior to actual inspections. This process hastened final acceptance which allowed the owner's building operations to open early. An additional crew was formed for the purpose of creating and completing the final punch list. This approach enabled the punch operations to start sooner than customary practice by overlapping this activity with the actual crews who were focused on the build out construction. The net result created the effectiveness of doubling personnel on each floor, which facilitated faster completion. It is also important to understand that the punch list process consists of multiple smaller duration activities as compared to build out construction. The differences in the rhythm of build out and punch list work processes has a tendency to be counterproductive when attempted by a singular crew. A descriptive analogy would be the difference in technique when running in a marathon (build out) versus a sprint (punch-out).

Lastly, contingency planning was implemented for the installation of the MRI units as the owner wanted to take advantage of the latest technological advances in the equipment. As a result, the contractor prepared a contingency plan by designating a pathway on the 18th floor of the building and creating an exterior skin panel assembly which could be easily removed to insert the MRI after the majority of construction was complete. This contingency plan required analysis of floor weight capacity, hallway width, and exterior modular assembly.

3.5.3. Budget Control Risk Mitigation

Pre-procurement planning as a means to control the budget was implemented for the building crown, aka Tiara, and the exterior skin. These components were separated in the design phase so that just-in-time decisions could be made on cost solutions thereby managing escalation risk.

> **Smaller Increments of Work creates Greater Management Capability**

The contractor mitigated escalation efforts through pre-procurement of high-cost/escalation items such as copper, sheet metal, and other commodity-driven products. This required coordination between the designers, contractor's estimators, and subcontractors to pre-procure materials and store them prior to the required site delivery dates. This concept required that quantities be identified as early as possible during the redesign efforts, so the owner could decide on the best possible date to authorize procurement. While quantities would vary somewhat during the scope modification phase of the project, the goal was to get as close to the quantity as possible so that the major risk of escalation was minimized.

3.6. Practices

3.6.1. Leadership Consistency Practices

In addition to mitigation measures, several practices were employed by the team in an effort to ensure high performance. Leadership practices can have the most stabilizing effect on successful project execution. Specific means and methods implemented on the project in this case study are illuminated as follows. The first was a focus on timely decision-making by all members of the team. The only exception to this tenet was with regard to the medical equipment and in this case, the decision was deliberate to delay until the very last moment in order to provide the most recent technological advances.

> **Timely Decision-making is Crucial to Schedule Maintenance**

Secondly, the basis for decision-making was consistent and clearly communicated to the entire team, "the budget and schedule were to be maintained and any request or condition that transitioned beyond those parameters were to be communicated to ownership immediately." While there was turnover in key leadership positions from ownership, the timing was early in the project during the design phase. In addition, the leadership from the program manager, architect, and contractor remained unchanged during the construction implementation phase. The solidarity and singular vision between team leadership provided

> **Leadership Stability Cannot Be Underestimated**

the resolve necessary to work through the design modifications and the typically late medical procurement.

3.6.2. Scope Change and Time Management Practices

Scope modification and time management practices were also implemented to enhance production of the project. Multiple teams on both the design and construction teams enabled responsiveness to a project with constant change. Production analysis allows for adjustments during execution prior to time depletion of the task. This type of analysis aids the team through predictive forecasting and establishment of the most efficient means to complete a work component. The

> **Responsiveness to Scope Change is Key to Success**

net result can also produce a better rhythm during the work cycle which fosters worker engagement, satisfaction, and greater control during execution. Another effective means of management is to break down activities into incremental elements. Smaller details allow production personnel to more accurately reflect on the sequencing and timing required for successful completion. Lastly, prefabrication of repeatable building components creates the opportunity for more efficient execution in a controlled environment, offsite from the project. Installation of the prefabricated elements typically requires less time on-site which results in schedule float creation and allows accessibility of the work space by multiple trades.

3.6.3. Budget Control Practices

Pre-procurement planning practices were utilized to minimize the effects of cost escalation. This endeavor commenced by pinpointing those commodities that were the most volatile, specifically in this case, copper, sheet metal, and drywall. Since the design occurred concurrently with construction, the design and construction team had to project estimated quantities before the design was complete. Once a quantity was mutually agreed upon, the materials were purchased and stored until needed on-site. This process provided financial savings to the cost while simultaneously doubled the handling of materials and required more planning for delivery.

3.6.4. Environmental Risk Practices

Lastly, an emergency preparedness drill was implemented to determine the most critical areas which needed protection and the overall timing for execution of the measures. This knowledge allowed the team to forecast the latest that work could continue prior to an environmental event. Confidence in the ability of the project site to withstand damages due to wind and rain also facilitated shorter recovery times.

3.7. Final Outcome and Benefits

During separate interviews with the project team members (owner, architect, and contractor) each was asked to describe the tangible and intangible benefits which resulted from the mitigation measures.

3.7.1. Tangibles

Tangible outcomes from the owner's perspective was that a more aptly suited building was delivered to meet the end user's purpose, and flexibility for growth was enhanced with the addition of future expansion space. Since the building's opening in July of 2010, the expansion space occupancy rate has grown to 90% of total allowable square footage. Shell space management created opportunities to consolidate services for one of the largest departments, orthopedic medicine, which resulted in business efficiency and increased customer satisfaction due to convenience. The architect's perception reinforced the expanded usage capacity of the building space and the ability to adjust to meet the changing needs of the institution. From a construction execution perspective, the contractor believed the emergency preparedness planning yielded shorter periods of downtime required for damage repairs after significant environmental events. Both the contractor and architect acknowledged that more work was compressed within a fixed period of time and that this couldn't have been possible without the dual team approach implemented by the designers. Net results produced a completed project five weeks earlier than planned and under budget. Both the program manager and contractor attribute this success to the construction mitigation efforts described in this chapter as well as the early planning efforts for activation by the health care operators.

3.7.2. Intangibles

From an intangible perspective, the contractor voiced the most sustaining benefit to their company culture, as compared to other team members. The pre-planning activities proved to be so successful in improving performance and timing that the company incorporated many of the new processes as part of their standard operational procedure for future projects. A secondary intangible benefit from incorporation of new practices was the creation of a learning culture amongst the staff members. This new cultural caveat was supported through "Lessons Learned" evaluations within the model for building future health care facilities. The practice of questioning how installation of specific building components could be improved and through what means instilled intellectual curiosity from the construction team, subsequently providing proactive risk assessment during the planning and execution phases of a project. This pre-emptive approach creates risk management strategies which by their nature facilitate contingency planning thereby improving the probability of timely completion.

The owner, architect, and contractor shared the similar opinion that the creation of pre-core team meetings facilitated the ability of leadership to ensure a positive work environment in spite of adversity and difficult schedule parameters. Additionally, the structure of a smaller leadership group allowed timely decision-making which was critical to project execution since a multitude of concurrent design and construction activities were performed. The bonus of this practice was the improved relationship of all team members both during and post-construction phases. The project was completed 7 years ago and the leadership members from each firm still share a learning culture with one another while working on separate projects.

3.8. Summation

The key contingency planning and mitigation efforts are summarized in Figure 19. Each team member's involvement is illustrated by the following color designations: Architect (yellow), owner (green), and contractor (red). The primary principles driving all team members were singular vision, leadership cohesion, and timely decision-making which was facilitated through the implementation of pre-core team meetings. These meetings were characterized with participants as the leaders of each firm, utilizing data collected from the larger core team meetings, for the purpose of summarizing key issues so that decisions meet the projects timing needs. The vertical columns from Figure 19 represent the practices employed by each team member which successfully contributed to early schedule completion of the project.

- Separate building permits by floor for faster closeout of work with regulatory inspectors.
- Early procurement of materials which were impacted by cost escalation.
- Implementation of multiple design teams to accommodate redesign and construction operations created by the reprogramming of the building.
- Early constructability planning of major systems which facilitated more time-efficient methods of construction.
- Owner and contractor early occupancy planning about a year prior to move-in.
- Production and trending analysis of construction systems which allowed for time adjustments to stay on schedule.
- Utilization of nonproduction hours for material stocking in the construction work space.
- Prefabrication of major building systems which reduced jobsite installation durations.

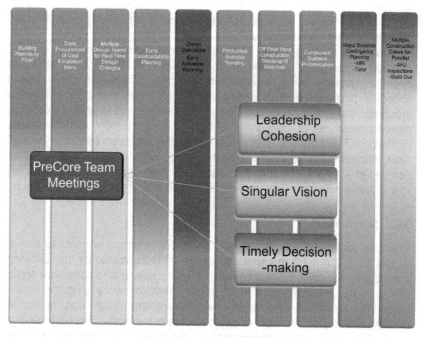

Figure 19:   Case Study 1: Contingency and Mitigation Efforts by
Team Members. (Architect, Yellow; Owner, Green; Contractor, Red)

- Contingency planning for major systems, i.e., MRI's and building crown, so that complexity issues wouldn't impact other system installations.
- Implementation of multiple construction crews aligned to work flow, i.e., punch list crew, inspection crews; which are short term processes vs the typical installation crews.

Based upon the overall duration of the project, each of these practices created opportunities for schedule compression.

From the owner's perspective, strategic business-driven issues were late in the design and construction process which prohibited a considerable amount of contingency planning. While the owner recognized that the initial decision to reprogram the building created a difficult situation, the process enabled a better building to meet the needs of the users resulting in a successful project. The contractor's overall synopsis was that the schedule was so dynamic with delays induced by the late stage design modifications, unforeseen foundation obstructions, and environmental events, that a contingency plan was difficult to develop proactively. As a result, project success was dependent upon continuous mitigation efforts and efficiency in construction production.

Case Study #2: Team Integration, Decision-Making, and Schedule Reporting, are Critical Factors for Scheduling Success

3.9. Project Scope

The project is a medical hospital located within the Medical Center, in San Antonio Texas and consists of 10 floors totaling 1 million square feet. The scope of the project was new construction in an urban setting. The cost of construction was $770 million with a commencement date in 2011. Project before and after photos are shown in Figures 20 and 21, respectively.

3.10. Key Project Milestone Dates

- Architect hired by institution: Summer 2009
- Program manager hired by institution: Spring 2009
- Contractor hired by institution: Fall 2009
- Key subcontractors hired by contractor: March 2010
- Original construction commencement date: February 2011
- Owner activation period:

Planning commencement date: March 2012

Move commencement date: February 2014

Actual first patient (projected date May 24, 2014)   April 11, 2014

Figure 20: Project "Before" Photo.

Figure 21:   Project "After" Photo.

3.11. Practitioner Team Members

Bexar County (owner) provides health care services to the San Antonio Region and was established in 1917 with an annual number of inpatient days more than 158,000 as of 2015. The architect, established for 82 years, provides design services on an international level with $484 million in revenue (Dodge Data Analytics, 2015a, 2015b). Relative to timing, it is important to understand that the architect worked on the project only three months prior to the addition of the contractor. The program manager was the first member of the team procured by the owner. The program manager was a joint venture between two firms (Broaddus & Associates, 2009). One firm is characterized as a regional health care manager while the other is local to the San Antonio area with a minority business enterprise certification (Dodge Data Analytics, 2015a, 2015b). Their combined history spans 107 years. The general contractor for the project was a joint venture comprised of three firms. The cumulative experience of the contractor team spanned 201 years. Areas served by the construction firms ranged from regional to national markets. All firms were included in planning and execution efforts for the smaller enabling projects. These projects were necessary to restore hospital functions from areas which were removed to make ready for the new hospital facility. The design assist delivery method was utilized for early procurement of key subcontractors (mechanical, electrical, plumbing, fire sprinkler, and low voltage). This method allowed the major subcontractors to work with the team during both the planning and execution phases of all projects.

3.12. Situation and Risks

As illustrated in the project timeline, the team was selected within a six-month span. This timing allowed the team a little more than a year to work together during preconstruction. By the time, the structural construction reached 49% of completion, the following schedule delays had occurred:

- Scope change to foundation piers due to unforeseen conditions impacting the pier elevations.
- A stop work order had been issued on the catheterization laboratory due to design changes.
- Unforeseen conditions for the water line through the site had been encountered.
- Additional excavation scope to the basement had been added due to scope change.
- Excessive weather impacts had occurred during winter months due to ice and rain. It's important to note that these weather conditions are not prototypical to the region. The net result of the delays was that all weather days had been used for that period and an additional 45 delay days had impacted the critical path.

All of the issues enumerated above, resulted in The JV contractor claiming a 14-day extension for time plus general conditions cost of $15,000 per day. Additionally, an acceleration plan had been submitted to the owner to mitigate all but two days of the delayed time in the amount of $45,000 per day. Acceptance of the acceleration measures would essentially mitigate the schedule losses but would cost approximately $500,000.

Upon receipt of the acceleration plan, the owner commissioned an independent third-party review of the schedule. At the time of this review, the overall project was 29% complete with the building structure 49% complete. The lapsed time of the contract was at 32% so construction was approximately 3% behind schedule (Figure 22). The activity count had increased by 7% from the baseline schedule which reflected changes in

Progress of Work Reflected in Construction Schedule	
Start 25-Feb-11 Finish 24-Apr-14	
Overall Schedule Completion	29 %
Site work Completion	0 %
Structure Completion	49 %
Baseline Schedule Activity Count Nov. 2011 –	9087
Current Schedule Activity Count Feb. 2012 –	9736

Figure 22:   Project Status Report.

scope. The purpose of the review was to assess the acceleration plan and determine the feasibility of the remaining schedule to meet the desired deadlines. As with any endeavor, the intensity required during challenging projects can narrow the teams focus. Oftentimes, the team benefits from a broader perspective provided by an external agent. During the collaborative work sessions between the team and the analyst, several issues became apparent. The first of which was insufficient communication from the team to the client with regard to potential schedule impacts. This condition contributed to reactive decisions and a sense of frustration by the team. The team expressed the need to become more integrated and proactive in their planning and decision-making. The secondary issue revolved around how to organize a large team to facilitate efficiency in strategic forecasting and decision-making. It's important to remember that the team was comprised of six firms (architect, contractor, and program manager). One of the inherent risks with a team this size is inefficient communication flow to the leadership level.

Another common situation existed, in that the architect was still in design mode for scope changes on the project (Figure 23). As noted in the timeline, all team members were assembled within a six-month span. This is advantageous in terms of achieving alignment between all team members, however, the architect is at a disadvantage since design for a

Package	Pages	Date	Description
TW-1	19	11/30/10	TW(-1) Excavation/Piers
AD#1	18	12/21/10	Civil/Structure to TW-1
PR#1	29	1/25/11	Pier Caps
PR#2	54	3/1/11	Underground MEP
PR#3	36	3/17/11	Underground MEP and Underfloor Grading
PR#4	13	4/12/11	Pier and Crawlspace Grading
PR#5	222	4/14/11	TW(- 2) Superstructure
PR#5R	247	4/29/11	Addendum to TW-2 Superstructure
PR#6	20	5/27/11	Underfloor MEP
PR#7	149	6/10/11	Underfloor MEP & Elevator Oil Seperator
ASI#1		6/18/11	Underfloor Plumbing
PR#8	51	7/7/11	HVAC Piping in Crawl Space/ Areaway modif
PR#9	1777	7/21/11	Tower - Interior Finishes - TW-3
PR#10	369	8/11/11	Interior Finishes(AD#1-TW-3)
PR#11	198	8/25/11	Interior Finishes(AD#2-TW-3)
PR#12	1128	9/12/11	Supplemental and RFI's
PR#13	3	10/24/11	Supplemental and RFI's
PR#14	377	11/10/11	Structural/Architectural
PR#15	450	11/14/11	MEP/IT
PR#16	104	12/7/11	Supplemental and RFI's
PR#17	599	1/26/12	Supplemental and RFI's
PR#18	49	1/26/12	Signage Package
PR#19	430	2/23/12	Supplemental and RFI's

Figure 23: Project Design Plan Changes.

project of this magnitude typically spans a year or more. This condition created a reactionary affect by the contractor to adjust construction to maintain the construction schedule. It is common that most service providers (architect, program manager, and contractor) find it difficult to say "No" to the client, even when it presents difficulty to perform their jobs. After all, the primary job is to provide a physical structure for the owner's operational vision.

Lastly, environmental issues had impacted the schedule with excessive weather at a time when the project is most susceptible. To compound the issue, the structural work was only half complete so the potential for additional impact was probable. In this case, the term environmental encompasses more than weather impacts. The work conditions became a factor due to the increased number of workers due to acceleration. As is typical in this condition, safety becomes a concern since the work space effectively becomes smaller as more workers are added. Increased work hours by crews also require close monitoring since the workers are prone to become tired over extended periods of long hours. Mobility to and from the work zones in a multistory building can also impact the schedule when the worker count becomes elevated. Perhaps, the most important question becomes, how can the worker environment incur the least impact from uncontrollable weather conditions?

The challenges described above can be summarized in four categories: team integration, decision-making protocol, schedule reporting at the executive level, and environmental. The specific methods employed by the project team are chronicled below and offer specific ideas for time mitigation and contingency management.

3.13. Mitigation Efforts

The mitigation efforts have been categorized by the associated risks in order to provide easy reference by the reader. As defined in the previous section, the major risk categories are team integration, decision-making protocol, schedule reporting at the executive level, and environmental impacts. The mitigation measures are discussed in the following paragraphs and summarized in Table 3.

3.13.1. Team Risk with Integration Mitigation

Upon completion of the third-party review, the project team established a more integrated structure. The assimilation efforts included formation of specific functional teams which focused on proactive schedule reporting and activation. Each team was a sub group to leadership (aka. pre-core team) and included personnel from each firm. The reports for each subteam funneled upward to the pre-core team. While each team would

Table 3: Project Mitigation Efforts.

Risk	Mitigation Technique
Team integration	• Collaborative third-party work session • Establishment of internal teams for specific purposes (i.e. Activation, Proactive schedule analysis) • Expedited punch list documentation and approvals through use of technology
Decision-making	• Create culture of decision management by forecasting decisions required within the schedule • Establish regular executive meetings and processes for project impact decision-making • Establish expectations and the life cycle for decision-making • Early activation planning • Cessation of design changes to allow focus on production activities • Analysis of current production to industry standards to shape the range of decisions to be made
Schedule reporting at the executive level	• Standardized report in summary format • Established cycle of reporting and review • Incorporation of all key elements into the schedule (i.e., medical equipment, owner decisions, outstanding architectural changes)
Environmental impacts	• Acceleration plan for the structure (six days a week) • Temporary dry-in on Levels 5 & 12 • Re-sequenced installation of HVAC units to facilitate earlier riser flushing • Doubled crews for millwork and flooring to facilitate faster installation • Installation of MEP overhead between re-shores at the basement level • Analysis of connection points between the existing hospital and the new facility

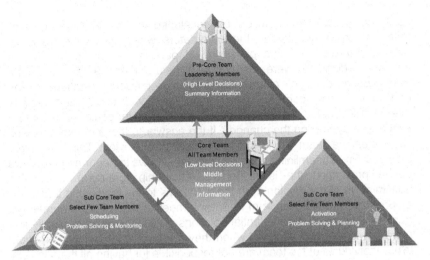

Figure 24:   Team Structure.

evaluate information on a microlevel, the reporting goal was to forecast macrostrategies. By reporting the "big picture" to the leadership team, strategic decisions could be focused on faster. This approach assisted the overall team to remain on schedule. The scheduling subteam included the scheduler and operational staff responsible for driving the schedule. The frequency of the meetings was determined by the intensity level of the project. For example: during high-risk potential impact periods, that is, vulnerability to weather, the team convened weekly. The meeting frequency would vary to bi-monthly during low-risk periods. An additional team was formed for activation planning of the building. This subteam also reported to the pre-core leadership team and was established with the specific goal to reduce the activation period. Reduction of this phase would result in a shortening of the project's critical path or overall duration. Team integration was also facilitated through shared use of software to generate and complete the punch list. Both the architect and contractor used this system, which saved a considerable amount of time in the process of readying the space for occupancy. Team alignment with the strategy to proactively address timing issues resulted in early completion (Figure 24). This integrated focus cannot be understated as to its impact on later mitigation efforts and team environment.

3.13.2.   *Decision-Making Risk Mitigation*

With the emphasis on forecasting and collaboration, the natural next step was to focus on establishing a culture of prompt decision-making. In order to capitalize on proactive planning, decisions must be made timely. Culture creation is facilitated by communicating clear expectations and protocols, so it was critical that protocols be developed for decision-making. If all team members understand the parameters which guide the decision-making process, then the possibility of paralysis is minimized. As an example, if decisions were cost neutral in nature, then the next parameter to be applied does the change facilitate more efficient operations by the end user. If the decision would incur additional costs, and provide operational benefits, then the differentiating parameter could be "is there a cost return by making the change." If the answer to those parameters was "no," then the team would know to reject the decision. This framework provides time streamlining because the team knows to perform the analysis before presenting the issue to the leadership (pre-core) team. The transmission of information from the subteams to the leadership team also emphasizes the importance of narrowing meeting focus for the purpose of framing issues and their potential solutions. Team members generally feel more productive and engaged when they are involved in solution creation. While individuals of the team are not responsible for having all the answers, an effective team can create multiple options. To maintain this environment, it is essential for meeting cycles to be consistent. Mundane activities should be encouraged handled outside of the meetings. Essentially, energy creates more energy, so the leader of each team should evaluate the team's engagement and attentiveness to problem-solving.

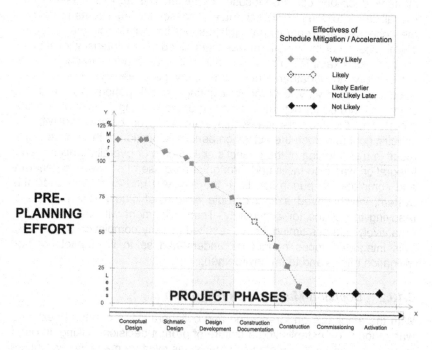

On this project, this approach led to several very important decisions which positively aided the overall schedule. First, the decision was made to cease all future design changes to allow the construction execution team to focus on completing the work before the scheduled deadline. This may sound like an easy decision, but from a supplier's perspective, client satisfaction is always important and saying "no" to desired changes is difficult. The decision to stop design changes must have buy in from the owner, architect, program manager, and contractor to be effective.

The second major decision was to proactively investigate the mechanical, electrical, plumbing, fire sprinkler, and structural tie-in conditions between the existing and new structure. This included

> **What You Can't See Will Always Hurt You!**

utilities located within ceiling cavities as well as structural attachments. Investigative minimal demolition was implemented to determine if conditions were as drawn. The purpose for these actions was to confirm existing and new utilities interface prior to construction commencement. The strategic key to this decision is that oftentimes existing conditions are different than originally anticipated. Early discovery of the hidden conditions allows for adjustment in constructability before they impact the overall schedule.

Another key decision made during activation planning was to determine the priority operational areas. Figure 25 illustrates the priority functional area on a floor. By identifying these areas early during construction, the contractor had the ability to adjust work crews to accommodate this objective. This also served to focus the end user in their equipment and move planning. The original occupancy/move period was originally projected for a three-month duration. This period was reduced to six weeks. Contributing factors included 23 months of planning, temporary certificates of occupancy granted in key areas for early occupancy, and prompt closeout through collaborative technology. The last decision involved commissioning an independent third party to analyze the production rate of construction to a comparably sized project. This decision was requested by the owner and supported by the contractor. The comparison project was similar in total square footage and structural composition. Other similarities included construction of a central plant, parking garage, and a medical facility. The main difference between the projects were the building height, 10 vs 25 floors. The floor height delta dictated that a comparison of completion dates by floor would not be equitable, but the completion on a square foot basis was reasonable to evaluate. Refer to Figure 26 for the analysis chart between the two projects. The significance of the comparison exercise was the realization that productivity per square foot was relatively similar between both projects. This finding provided two important realizations for the project team. The first was a sense of confidence in their ability to execute construction at a productive rate. With this understanding, the team

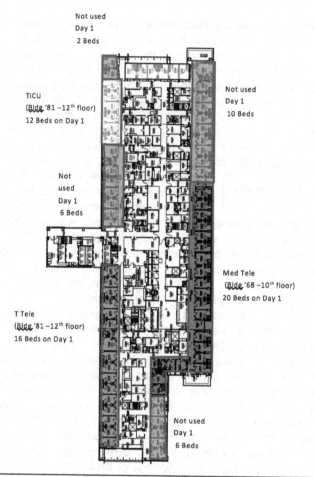

Not used
Day 1
2 Beds

TICU
(Bldg '81 –12th floor)
12 Beds on Day 1

Not used
Day 1
10 Beds

Not
used
Day 1
6 Beds

Med Tele
(Bldg '68 –10th floor)
20 Beds on Day 1

T Tele
(Bldg '81 –12th floor)
16 Beds on Day 1

Not used
Day 1
6 Beds

9th Floor – Sky Tower –Transplant, Cardio-Thoracic, Cardio

Figure 25:   Project Floor Plan.

validated their baseline performance on completed activities. The second realization framed future activities. Since the baseline production was comparable to industry standards, the team could now focus on strategies to minimize future activity durations. This was important to the owner since overall schedule success included performance of the activation phase. Hence, the decision was made to start activation planning early with the goal of shortening the duration and finish the project earlier than anticipated.

Level	Elevated SF	Start	Finish	Duration (Cal.) w/o Weather	WKS	Productivity sf/day	Effective Duration
Buildout (EXCLUDES 11th floor)	892,003	5-Jun-12	29-Jan-14	603	86	1479	
Skin to Dry-in	947,346	6-Apr-12	18-Feb-13	318	45	2979	
Dry In	947,346	7-Mar-11	18-Feb-13	714	102	1327	
Conditioned Air	947,346	7-Mar-11	21-Dec-12	655	94	1446	
MEP Systems Start-Up	947,346	7-Mar-11	18-Dec-12	652	93	1453	
Permanent Power On Line	947,346	7-Mar-11	12-Dec-12	646	92	1466	
Central Plant (Less Phase 2)	47,800	16-Jan-12	3-Dec-12	322	46	148	
Elevators	947,346	7-Nov-12	29-Jan-14	448	64	2115	
Substantial Completion	947,346	7-Mar-11	21-Feb-14	1082	155	876	
TCO (Podium Levels)	947,346	7-Mar-11	3-Dec-13	1002	143	945	
TCO (Tower Levels)	947,346	7-Mar-11	27-Jan-14	1057	151	896	
CO	947,346	7-Mar-11	27-Jan-14	1057	151	896	
1st Patient PER GC SCHEDULE	947,346	7-Mar-11	21-Feb-14	1082	155	876	
Final Completion	947,346	7-Mar-11	24-Apr-14	1144	163	828	

This date is predicated on same span of OPC of 59 days between SC and 1st Patient (not recommended until transition planning is

Represents Effective working time

Production rate is skewed due to OPC serving 25 floors and UH only 10 floors

Good Productivity Rates

Difference in Exterior Skin productivity is based on precast exterior OPC vs. Brick UHS

Figure 26:   Third-party Productivity Analysis.

3.13.3. *Risk Mitigation through Schedule Reporting*

Now that the team had dealt with risks associated with team integration and decision-making, the natural progression was to ensure effective communication. Regardless of the project, the one risk that can derail success is communication between the production personnel and the decision-makers. Experience has repeatedly shown that to effectively communicate with leadership the contractor must remove the technical jargon. Contractors are naturally drawn to the technical side of the building business; however, most clients are focused on their operational business models, not construction. An example of this occurred at lunch with a program director from a medical provider (client). The contractor had just been awarded a large project. Discussion during lunch began to delve into the complexity of the size of the piers associated with the foundation work. After several minutes of talking, the contractor realized that the client had no understanding of the technicalities being discussed. The result was that the client felt less than smart and did not care how the work was to be performed. In their view, the important thing was that the contractor fulfilled the obligation of providing the building so that the core business could commence on time. It became painfully clear that it was important to speak and write in a shared language by the owner and contractor. That commonality is known as risk. The other relative communication tool is to frame decisions by the risk and the means to mitigate. And lastly, ensure that the information can be presented in a summarized format.

Monthly Schedule Executive Summary
Reporting Period: March 2012

Summary Dates (Data Date: 3/31/2012)

	Contractual	Current	Var. (Cal. Days)
▶ Tower Podium Ready for Staff/Stock	N/A	12/3/2013	
▶ Tower Substantial Completion	1/16/2014	1/30/2014	-14
▶ Tower 10th Floor Substantial Completion	2/13/2014	2/21/2014	-7
▶ Tower Ready for CUP Utilities	N/A	11/9/2012	
▶ CUP Utilities Available to Tower	N/A	12/20/2012	
▶ Tower MUST HAVE CUP Utilities	N/A	1/15/2013	
▶ CUP Substantial Completion	1/11/2013	1/24/2013	-13

Figure 27: Executive Summary Schedule Variance.

Presentation is critical at the leadership level for two reasons. The first of which is that leaders are involved in multiple endeavors, so their availability can be limited to the execution team. Secondly, the decision and risks must be honed to their simplest level when it reaches leadership. This ensures that the team has thoroughly worked through the problem. When working with the leadership team, there is nothing worse than presenting a half-developed risk solution for review. This wastes valuable time and erodes confidence away from the production team. All of these factors were contributors to the Executive Schedule Summary Report, Figure 27. A comprehensive report includes the following reoccurring elements:

- Summary of current progress with regard to projected substantial completion vs contractual substantial completion.
- Identification of potential schedule impacts and proposed methods of mitigation
- Documentation of actual mitigation progress (Figure 28).

Actual Mitigation of Delays to the Schedule:
1. **Tower – Conditional Permit Expires on April 2, 2012**
   Conditional Permit has been extended to May 4 and the hold on MEP inspections was waived. No Impact to Critical Path.

2. **Tower – Waterproofing Detail Delaying Scaffolding Added for Mitigation**
   ZVL accelerated the scaffolding. No Impact to Critical Path.

3. **Tower – Waterproofing at EC Deck Delayed by Rain**
   Existing float was used to mitigate the delays. No Impact to Critical Path.

Potential Pending Delays to the Schedule:
1. **Tower - Evaluation of PRs 17, 19, & 20 in Last Month**
   Pricing has been turned in for PR 17. Schedule impacts are still being evaluated.

2. **Tower & CUP - Smoke Compartment per COSA CMR at Emergency Level**
   The meeting with COSA (April 2, 2012) resulted in at least the following modifications: 1) added fire smoke dampers; 2) two-hour duct wrap on existing ducts; 3) above ceiling fire (alarm or sprinkler, per COSA) modifications; 4) fireproofing of existing steel structure. This work needs to be completed by May 1, 2012 to avoid impacts to the critical path. ZVL is evaluating other mitigation options in the event this date is not met.

Figure 28: Executive Summary Schedule Mitigation.

- Identification of significant decision dates from the owner within the next couple of months.
- Documentation of weather days with identification of those which impact the critical path.
- Document requested and approved schedule extensions (Figure 29).
- Content inclusion of any item which represents a risk to the schedule (i.e., equipment furnished by owner, pending architectural changes).

Proposed Methods to Mitigate Delays to the Schedule:

1. **Tower Recovery Plan** – Delays from November 7, 2011 to February 29, 2012
   ZVL currently working with project trade contractors on a recovery plan, to include cost estimates, for review at the end of April 2012.

2. **CUP Recovery Plan**
   ZVL is working on a revised baseline to incorporate the impacts from the soil contamination and additional sequence changes since the August 2011 baseline.

Awarded Delays to ZVL's Contract:

1. None since last reporting period

Scheduling Process Update:

No changes to scheduling process since last reporting period.

Figure 29: Executive Summary Schedule Extensions.

The report should be reviewed immediately after publication with the leadership team. The cycle of report production and review should be consistent. The test as to whether the report is effective will be evident in subsequent discussion by the owner with the production team about projected risks. If this occurs, then it is a signal that the communication gap has been bridged successfully.

3.13.4. *Environmental Risk Mitigation*

The last project risk relates to environmental issues, some of which are not ecological in nature. The environment can also encompass conditions or surroundings. In the next illustrations, the working conditions of the building were affected by the mitigation measures. In some instances, the influences were positive while others created environmental challenges. The initial mitigation action involved sequencing the concrete forming operations from Level 5 to 10. While this effort failed to save time on the critical path, the structure was finished 11 days early. The critical path shifted to include the 4th-floor air handlers in lieu of the 11th-floor air handlers. The subsequent aspect of the modification was provision of conditioned air sooner than originally scheduled. This facilitated a better environment for the workforce and earlier installation of humidity affected construction materials.

Implementation of the acceleration plan required a six-day work week. While this was effective in mitigating a schedule delay, the work environment was susceptible to increased safety risks. In this particular case, minor safety incidences had increased during the acceleration period. As a result, the contractor assigned additional safety personnel and safety incidences dropped. The significance of this situation is understanding the correlation of time to the work environment. Changes associated with time acceleration often have a ripple effect on the environment and should be analyzed prior to commencement.

Temporary dry-in was executed on levels 5 and 12 to enable the work crews to commence interior construction earlier than planned. This method impacted the environment in that different trade work crews were able to spread out. This maneuver creates a more productive environment by providing working space and avoiding conflict between the trades. For example: the first trades to work in completed structural space are typically drywall (D), mechanical (M), electrical (E), plumbing (P), and fire protection (F). All trades commence work in the overhead space of the building. Conflict is created when they all start at the same time. Typical good practice is to phase installation start dates in the order of; drywall metal stud ceiling track, mechanical ductwork, plumbing service lines, electrical distribution conduit, and fire sprinkler piping. In summary, mitigation measures should consider the impact on the workforce prior to implementation. Productivity is a large factor in any acceleration or mitigation plan.

In addition to the increased locations for commencing DMEPF overhead rough in, the construction team calculated a plan to install the work between the re-shores of the structure in the basement. This method is significant as it is not typically implemented due to the difficulty in creating a path large enough for the work crews between shoring. The process of shoring and reshoring is illustrated in Figure 30 below.

The impact to the environment is reduction of working space and maneuvering of equipment. While this effort created an overall reduction in duration, there would have been some loss of productivity due to the cramped work area.

The largest time saving measure was to utilize doubled crews for millwork and flooring. After discussing the productivity is reduced by introducing too many workers in one area, these may seem counter intuitive. However, the floor square footage was quite large at 100,000. By comparison to Case Study #1 this project was almost three times larger. The massiveness of the floor space allowed millwork crews to start on one side of the floor (east), and the flooring crew started on the other side (west). Ultimately, the two trades would momentarily cross like "two ships in the night" at the center of the floor. This would minimize cramping the space for conflicting work crews. Or, before the intersection of crews occurred, one of the trades, could pick up and move to where the other

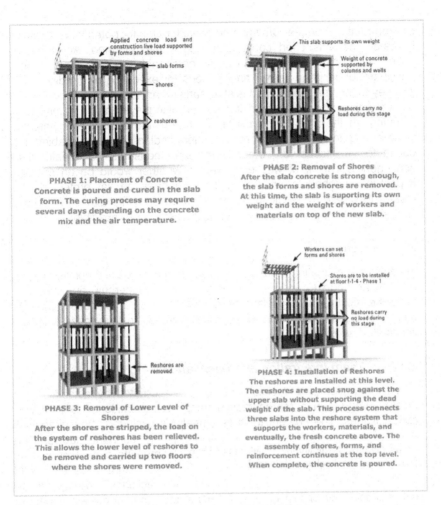

Figure 30:   Shoring Methodology.

trade had started, and the intersection would be completely avoided. This technique can be engaged on smaller buildings, but the practitioner must be aware that as the floor space becomes smaller so does the productivity factor.

The planning mitigation effort to analyze connection points between the existing hospital and new facility had the potential to disrupt hospital operations. This exercise involved some destructive testing of existing conditions, meaning that key areas would need to be demolished enough to expose hidden conditions. Care must be taken to ensure that the end users operational area be minimally disrupted and protected from environmental changes. Caution and consideration for potential disruptions indicated that this effort could take longer than planned. Anytime that coordination involves nonconstruction personnel, the scheduler should anticipate longer than normal time periods for execution of the discovery. The key to this mitigation effort is start earlier than later.

The typical environmental risk, weather, also played a role in mitigation planning. At the time of the third-party report, the exposure of potential weather impacts was 13 months. Understanding this risk enabled the contractor to focus on other mitigation measures to create float contingency. This float was essential as weather would be an ongoing impact to the schedule until the building interior space was protected. The means to create schedule contingency were primarily aimed on getting workers into the interior space construction as soon as possible.

3.14. Practices

In addition to mitigation measures, several practices were employed by the team in an effort to ensure high performance. Leadership practices can have the most stabilizing effect on successful project execution. Specific means and methods implemented on the project in this case study are illuminated as follows.

3.14.1. *Team Integration and Decision-Making*

Team integration and decision-making was facilitated by establishing a smaller group with a representative from each practitioner to deal with critical issues. The content for these meetings excluded "housekeeping" or mundane topics. The meeting cycle was weekly so that issues and decisions were made

> **Size Matters! Typically the Smaller the Group, the Faster the Decision-Making.**

either proactively or just in time. Perhaps, the most significant aspect of these meetings was that interruptions were not allowed, which enabled a focused effort on problem-solving. The approach toward meeting content was centered on risk management. The smaller setting created a more effective environment for direct communication. The interactive culture shifted when the team understood that speaking directly to the issues was the goal, without concern of political posturing or interpretation. Simply stated, trust between team members and empowerment to speak to issues was established during the third-party work sessions.

3.14.2. *Information Management and Reporting*

White papers became another tool utilized to summarize major issues in an executive summary format. This concise delivery method enabled executive management from the owner's side to understand the issue in a brief representation. Brevity and demonstration of impacts to the project are considered to the best formula

Common Language is Crucial for Effective Communication

for facilitating decisions, especially from nonpractitioners. While the owners project director understands construction, those that he/she reports to typically do not. Information must be delivered in the way the end reader/decision-maker can easily understand. In this way, the problem can be focused on in lieu of trying to understand the message. Contractors are notorious for reporting in "construction speak", also known as technical jargon. Oftentimes, the end user has no understanding of this language and do not want to take the time to learn. End users/owners' job is to make their business successful. The building is just a means to an end for their use.

Meeting cycles and standardized schedule reporting were maintained throughout the remainder of the project. Refer to Figure 24 for the meeting structure as previously described. The most significant reporting relating to maintaining and improving the project schedule was the Executive Schedule Summary (Figure 27, 28, and 29). This provided ease of understanding the schedule and efforts employed to manage project timing. Focus on forecasting key issues and decisions was elevated to the leadership level on a regular basis. This reporting technique also framed the remaining opportunities for time contingency creation. Regardless of the project size, this practice is recommended to translate the schedule from "contractor speak" to commonly understood risk management information (Refer to Appendix for Sample Report).

3.15. Final Outcome and Benefits

During separate interviews with the project team members (owner and contractor) each was asked to describe the tangible and intangible benefits which resulted from the mitigation measures. It was important to interview the firms separately so commentary would not become biased by group consensus. As practitioners generally prefer to focus on tangibles, these findings will be presented first.

3.15.1. Tangibles

From the owner's and contractor's perspective, the third-party work session was timely in meeting, the realignment needs of the team. In retrospect, the owner endorses an earlier schedule work session between team members and periodic reviews as the project develops. This external perspective illuminated issues that became diluted due to the everyday grind of the project.

Dealing with the major obstacle of lost time allowed the team to focus their efforts on anticipating major risk activities remaining in the schedule. Resolution was made possible by the owner's willingness to listen to the problem and contractor's transparency with the schedule. This led to a proactive approach directed to the activation schedule (Figure 31). Activation included:

- owner furnished, and contractor installed scope of work,
- relocation of existing equipment into the new building,
- new equipment and furniture,
- move planning by the end user from existing to new departmental space,
- planning operational changes by the end user as a result of the new facility,
- patient communications of the new facility by the end user, and
- time impact on existing operations impacted by new facility.

Department	Move Date	Move Start Time
EC Staff	March 31, 2014	2PM
Hartman	April 1, 2014	1pm
Neuro/Neuro Science	April 1, 2014	1pm
IPCU	April 2, 2014	8am
MICU/CICU/Medicine/Med Surg/Med Iso	April 2, 2014	8am
Ortho Trauma	April 4, 2014	8AM
UH Trauma	April 4, 2014	9AM
Volunteers Department	April 6, 2014	2PM
Transfer Services	April 7, 2014	8AM
Pediatrics	April 7, 2014	1PM
Gift Shop	April 8, 2014	8AM
Ortho Transplant/GYN/GYN Onc/Other Surgery	April 8, 2014	1PM
Materials Management Staff	April 9, 2014	8AM
UT Trauma Physicians – Group 1	April 9, 2014	9AM
Blood Bank – Staff	April 9, 2014	10AM
Air Life/UH Facilities	April 10, 2014	5AM
UT Trauma Physicians – Group 2	April 10, 2014	9AM
Med Tel/TICU/TTEL	April 10, 2014	1PM
Surgical Services Staff	April 10, 2014	5PM
Central Sterile Staff	April 10, 2014	6PM
Blood Bank Equipment Move 1	April 10, 2014	8AM
Surgical Transplant/Surgery/Transplant ICU	April 11, 2014	1pm
UT Trauma Physicians – Group 3	April 11, 2014	9AM
Blood Bank Equipment Move 2	April 13, 2014	9AM
Patient Admissions	April 13, 2014	2PM
Blood Bank	April 14, 2014	6AM
UT Trauma Physicians – Group 4	April 14, 2014	9AM
Patient Care Coordinators	April 15, 2014	8AM
Materials Management Supplies	April 17, 2014	8AM

Figure 31:   Activation Schedule.

Basically, any activity that allows the building to function to its intended purpose upon occupancy was included in activation planning. Both the owner and contractor agreed that this was a key to delivering the project on time. The activation phase actually created six weeks of float with the institution capitalizing on earlier opening and revenue generation.

Utilization of a schedule within a schedule allowed decisions by the owner to be included in activity logic. To be specific, decision milestones by the owner were added to the schedule. This facilitated a better understanding by the owner of the decisions that were needed to meet the contractor's deadlines. It also illustrated the invisible time needed to make decisions that could erode actual work activity durations.

Lastly, the contractor and architect streamlined the punch list process to facilitate completion of the activation process earlier than planned. Both practitioners used a common platform, BIM 360, to prepare and finalize the punch list process. Both parties used iPads in the field to create instant accessibility to the punch list. In addition, multiple crews were designated for the punch list operations vs construction completion. This maneuver allowed the punch list process to commence earlier thereby reducing time on the critical path.

3.15.2. Intangibles

What was originally designed to function as a schedule review, morphed into bringing the team together to deal with major issues. The actual schedule review facilitated modeling an optimized functionality of the team with collaboration as the focus. Critical-thinking and problem-solving interactions were reshaped by the practitioners. The team became more communicative about the schedule and the anticipated challenges. While the contractors brought potential solutions to the table, they did not wait until they had all the answers to bring the issue to the forefront. Dialogue between all members of the pre-core team (owner, program manager, and contractors) became more open. These efforts would not have been possible without the owner empowering the construction team to create a risk management culture.

Proactive schedule forecasting became the norm in lieu of reactive schedule management. The tool through which this was implemented was the Monthly Executive Schedule Report. Clarity of potential obstacles which might impede timely success was exposed earlier in the process. This proactive approach increased the team's ability to "course correct" thereby improving team confidence.

The owner and contractor shared the similar opinion that the creation of pre-core team meetings facilitated the ability of leadership to ensure a positive work environment in spite of adversity and difficult schedule parameters. Additionally, the structure of a smaller leadership group allowed timely decision-making. The bonus of this practice was the improved relationship of all team members both during and post-construction phases.

3.16. Summation

With the biggest risk being the massiveness of the project and multitude of practitioners, organizational structure of the team was critical. Adjustments to the team structure and reporting were an important component of schedule mitigation. These modifications created an environment of proactive problem solving and streamlined decision-making by the owner.

The key contingency planning and mitigation efforts are summarized in Figure 32. Each team member's involvement is illustrated by the following color designations: architect (yellow), owner (green), and contractor (red). The primary principles driving all team members were team integration, schedule reporting, and timely decision-making. Environmental issues were considered with each mitigation/acceleration measure. The vertical columns from Figure 32 represent the practices employed by each team member which successfully contributed to early schedule completion.

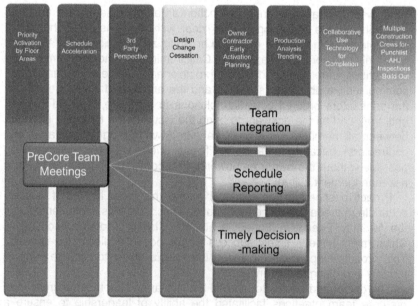

Figure 32:   Case Study 2 Contingency and Mitigation Efforts by Team Members. (Architect, Yellow; Owner, Green; Contractor, Red)

- Priorities were assigned to the operational units that needed to be open earlier. Temporary certificates of occupancy were utilized for faster closeout of work with regulatory inspectors.
- Schedule acceleration measures were implemented early in the project to maintain the schedule.
- External perspective to reframe the team's focus on creation of time contingencies.
- Decision to complete current work and cease design changes to ensure ability to meet the owner's project deadline.
- Early planning of end user activation to reduce the overall project schedule.
- Verification of production and baseline schedule through comparison to similar project.
- Collaborative use of technology by multiple team practitioners to complete the construction closeout more efficiently.
- Implementation of multiple construction crews aligned to work flow, i.e., punch list crew, inspection crews; which are short term processes vs the typical long-term installation crews.

The essence of this case study is, it is never too late to make adjustments. The team must be able to anticipate changing conditions and agile to adapt accordingly. When processes are in place to facilitate problem-solving and the team operates with a singular vision, success is within grasp of achievement.

Chapter 4

Risk Management Matrix

For students reading this text, the case studies in the previous chapter provide relevant context for the principles of change, risk, and mitigation. Experienced practitioners understand that learning from actual projects is always helpful, even if the project did not meet goals and expectations. Chapter 2 provided some valuable planning tools for risk and activity identification for project scheduling. This chapter is devoted solely to an instrument that has been developed by the author to identify areas of risk from multiple stakeholder perspectives. This tool is referred to as the Risk Management Matrix and can be found on the author's website at www.coblescorner.com. This risk identification tool is built such that individuals can complete the survey alone or in the presence of a group. For the process to realize maximum effectiveness, it is recommended that the project team collectively discuss the questionnaire answers to form a consensus understanding of each risk. By having a diverse representation of participants, owner, architect, engineer, contractor, maintenance personnel, end users, multiple perspectives of risk can be assessed. For example, the first question asks the respondents to rank the top four business units in the building project and rank them in importance to the business strategy of the owner. The purpose of this question is to discern the most important areas of the new project, so the schedule can reflect the urgency to complete this area prior to other spaces. This particular space may require additional activation planning, or should mitigation be required, the priority of completing this space is pre-established. The contractor may rank a space as more risk due to the complexity of construction. Regardless of the answer, a better plan can be developed when understanding is assimilated by all team members. Once this process is complete, the execution team will have guiding parameters with respect to priorities of completion. This can facilitate prompt decision-making as situations create challenges to completion.

The content of the questionnaire has been developed from experiences on many projects where risk factors had an impact on the project. The medium for this process is best served as a survey or questionnaire. This survey can be uploaded to a databank and pushed out to the respondents from an electronic platform. This format serves two purposes: the first is ease of use by the end user through an internet platform which can be accessed on any mobile device or computer. The second purpose is the ability of the database to collect all the answers to each question, so a cumulative risk assessment can be easily ascertained. The summarization of the risk level by question should be considered as a secondary data point as some risks from a specific practitioner may override the cumulative result from all practitioners.

The Risk Management Matrix is divided into two different areas of risk; planning and scheduling. The first sections of the survey address questions that relate to risk that must be addressed during the planning process. The second half of the survey deals with essential elements of risk that must be included in the project schedule as activities. For instance, a planning question asks, "Has the owner commenced contacting utility companies for necessary relocations, service additions, etc.?". The respondent is asked to answer if this process is well-defined, poorly defined, not defined, or NA to this project. If the answer was poorly or not defined, then the ranking would indicate high risk. Planning items which equate to a high risk, should be included in the project schedule during the preconstruction phase.

The Work Breakdown Structure (WBS) risk inquiries are directed toward content regarding construction, activation, inspection, and closeout phases. WBS is the organizational aspect of typical construction components in scheduling. The construction industry has relied upon this structure in schedules for many years and this framework is commonly accepted by practitioners. An example of one of the questions under the CONTRACTUAL REQUIREMENT SECTION asks, "have all owner approvals been indicated in the schedule?" This refers to the contract requirements for owner approvals during the design or construction phase of the project. The answer choices are "yes, no, partially, or NA." If the answer was "no," then the ranking for this aspect of the schedule would be a high risk. Each of the questions in the WBS portion of the survey act as a checklist to ensure that activities which would correspond to the risk are represented in the schedule. Upon completion of all the questions, the survey document will tally total risk numbers. The range of overall points would act as an indicator of how well the schedule addresses common risk elements to construction. As an example, if your survey scored a total point value of 200 risk points and the total possible high-risk points were 250, then it would be safe to assess that more work needs to be completed in the planning and scheduling of your project to minimize risk.

The timing of using this tool is intended to occur in the beginning of the project during the preconstruction period. By engaging early in the planning process, the likelihood of mitigating high risk items is greatest. The collaborative aspect of the instrument also reinforces the principle of team work. Variance in perspectives due to differing team member roles foster wider vision especially when challenges arise in the future.

The overall organization of the survey is outlined below (refer to Appendix for the entire survey document):

Part I: Risk Identification – Owner/End User Perspective
  • Establish understanding of owner business operations.
  • Definition of processes that are related to construction and involve the owner.
  • Identification of impacts to operations during construction.
Part II: Risk Identification – Design & Construction Perspective
  • As built conditions which can impact construction.

- Authorities having jurisdiction risk evaluation.
- Building configuration vs worker accessibility.
- Team composition – experience levels of team members.
- Identification of high risk project aspects.

Part III: Risk Identification – Schedule Elements
- Contractual requirements – WBS
- Design requirements – WBS
- Permitting requirements – WBS
- Owner procurement requirements – WBS
- Bid/pricing requirements – WBS
- Construction requirements – WBS
- Inspection requirements – WBS
- Activation requirements – WBS
- Closeout requirements – WBS

Each of these elements are important in proactively identifying and addressing risk in order to successfully achieve the desired completion date on a project. The following paragraphs will demonstrate reasoning as to their inclusion based upon four decades of construction experience.

4.1. Part I: Risk Identification – Owner/End User Perspective

Since all construction projects begin with the owner and end users, it is critical to get their perspective on the risks which may disrupt their ongoing operations (Figure 33). Oftentimes, this is the most overlooked aspect of construction scheduling due to the contractor's myopic focus on their needs to complete the project. While this section may be the smallest in terms of quantity of content, it can present the greatest intangible impact to any construction project. A key point of distinction needs to be made that this section is important regardless of whether the construction is adjacent or separated from an existing operational facility. The reasoning for this is, in the event that project delivery becomes in jeopardy, it could be important to focus on key operational areas to complete first so that partial occupancy can be achieved. A secondary reasoning for this focus is that activation time may take longer in a particular area of space which would require construction to complete sooner. The overriding point is that contractors are building to satisfy the purpose of the owner/end user. There are times when this goal becomes lost when singular vision is applied to technical building processes.

Section 3 – Establishing Understanding of Owner Business Operations attempts to identify the following for a specific project:

3.1. Top business units associated with the owner's business strategy (which are located in the new facility).
3.2. Top business units which are the most sensitive to noise disruption.

3. ESTABLISH UNDERSTANDING OF OWNER BUSINESS OPERATIONS (Completed by the Owner ONLY)				
Rank Importance to the Business Operations	Critically Important	Very Important	Important	Least Important
(1) Identify the top 4 business units in the building by importance to business strategy.				
a.	O	O	O	O
b.	O	O	O	O
c.	O	O	O	O
d.	O	O	O	O
(2) Identify top 4 areas which are most sensitive to noise and disruption.				
a.	O	O	O	O
b.	O	O	O	O
c.	O	O	O	O
d.	O	O	O	O
(3) Identify user groups that are impacted during construction and activation process.				
a.	O	O	O	O
b.	O	O	O	O
c.	O	O	O	O
d.	O	O	O	O
(4) Identify departmental service dates to the public.				
a.	O	O	O	O
b.	O	O	O	O
c.	O	O	O	O
d.	O	O	O	O
(5) Identify departmental activation periods for the dept.'s in item 4.				
a.	O	O	O	O
b.	O	O	O	O
c.	O	O	O	O
d.	O	O	O	O

Figure 33:   Risk Matrix Section 3.

3.3. The user groups which are impacted during the construction and activation process.

3.4. Dates for department/business unit's occupancy of the new space to meet operational goals, and

3.5. Anticipated activation period for each department/business unit.

The rankings created from the five inquiries listed above allow the building practitioners some understanding of the prime operational directive for each department within the new facility and the potential impact that construction can have on existing operations. This step facilitates owner engagement and understanding of how their new facility may impact them and begins communication between building practitioners and end users. This is significant since the process of constructing a new environment always begins and ends with the end users in mind.

Section 4 – Definition of Processes That are Related to Construction and Involve The Owner (Figure 34) – attempts to rank the following 22 processes and risk identifiers between the owner and contractor for a specific project:

4.1. Communication protocol between contractor and owner for an existing facility.

4. DEFINITION OF PROCESSES THAT ARE RELATED TO CONSTRUCTION AND INVOLVE THE OWNER	Not Defined	Poorly Defined	Well Defined	NA
(1) Has a communication protocol been established between contractor and owner for impacts to the existing facility during construction?	○	○	○	○
(2) Has a communication protocol been established to inform the end user of construction impacts to their operation?	○	○	○	○
(3) Have communication plans to the end user been developed for construction impacts to operational uses of existing building?				
(4) Has a process been established for utility shutdowns to existing systems for tie ins?	○	○	○	○
(5) Have the existing utility services been identified that will be impacted by construction?				
(6) Has the owner commenced contacting utility companies for necessary relocations, service additions, etc.?	○	○	○	○
(7) Has a room numbering protocol been defined for the purpose of identification of MEPF systems?	○	○	○	○
(8) Does the owner have a commissioning agent and process defined?	○	○	○	○
(9) Does the owner have an activation committee established for move in operations?	○	○	○	○
(10) Has the building maintenance department prepared a plan for systems management in the new building?	○	○	○	○
(11) Has site access been coordinated to determine pedestrian and vehicular pathways?	○	○	○	○
(12) Have all construction and operational inspections been identified?	○	○	○	○
(13) Have construction material storage and staging plans been developed?	○	○	○	○
(14) Has the scope of work been defined to constitute substantial completion been defined?	○	○	○	○
(15) Have roles and responsibilities been established for each practitioner team?	○	○	○	○
(16) Have communication protocols been established for the practitioner and owner team?	○	○	○	○
(17) Has a process for difficult issue problem solving been established? (i.e. smaller team, urgency period defined, parameters for decision making)?	○	○	○	○
(18) Was there collaboration by all execution teams in finalizing the schedule?	○	○	○	○
(19) Was the baseline schedule approved by the owner within 30 days of project start?	○	○	○	○
(20) Are their prequalification requirements for construction workers (i.e. security, background checks)?	○	○	○	○
(21) If the owner is self insured, is there a certification process requirement for contractors? And has that process been defined?	○	○	○	○
(22) Has the process for establishing a baseline schedule and documenting time extension requests been identified?	○	○	○	○

Figure 34: Risk Matrix Section 4.

4.2. Communication protocol between contractor and end user for construction impacts.

4.3. Communication protocol between construction practitioners.

4.4. Process for utility shutdowns and tie-ins.

4.5. Identification of existing utility services that will be impacted by construction.

4.6. Owner engagement with utility companies for the purpose of relocating services.

4.7. Room numbering protocol establishment for mechanical, electrical, plumbing, fire sprinkler (MEPF) system identification.

4.8. Commissioning agent engagement and plan development.

4.9.–4.10. Activation engagement by committee involving the end user, owner, building maintenance, and contractor.

4.11. Site accessibility impact study on pedestrian and vehicular transit during construction.

4.12. Identification of construction and operational inspections required due to construction.

4.13. Construction staging areas developed and coordinated with end users.

4.14. Identification of scope of work required to actualize substantial completion.

4.15. Identification of team members and their roles from owner, end user, all practitioner firms.

4.17. Process for problem-solving of difficult issues and identification of sub-teams to be involved.

4.18. Collaborative process in preparing the schedule by all stakeholders.

4.19. Approval granted of baseline schedule by owner within 30 days of project commencement.

4.20.−4.21. Process for construction worker clearance to perform, i.e., owner self-insured programs with additional credentials required by crew, and

4.22. Process for establishing a baseline schedule and documenting time extension request.

The matrix provides a ranking process of how well-defined each of the above elements are at the time of the evaluation. Each participant in the collaborative review session has the ability to rank each process or protocol from well-defined to not defined. The not-defined ranking constitutes higher risk in the planning stage of the project. Experience has shown that failure to address these issues during preconstruction creates a reactionary environment. Identifying these processes after construction commences can slow the problem-solving process, when time is at its most precious. The intangible factors of a disciplined decision-making, sense of urgency, and proactive forecasting are common elements for each of the questions in this section. Typically, a prepared team will facilitate smoother operations and incur less stress while executing the project.

Section 5 − Risk Ranking of Impact to Owner Operations During Construction (Figure 35) attempts to assess the current risk level of potential impacts per the following:

5. RISK RANKING OF IMPACT TO OWNER OPERATIONS DURING CONSTRUCTION	High Risk	Moderate Risk	Low Risk	NA
(1) Has the commissioning process been distributed to the contractor? (No=high risk)	O	O	O	O
(2) Has the activation plan been developed for end user move in? (No=high risk)	O	O	O	O
(3) Has equipment planning commenced by the owner? (No=high risk)	O	O	O	O
(4) Has the IT plan been developed by the owner's IT department and consultant? (No=high risk)	O	O	O	O
(5) Has the temporary wayfinding and signage plan been developed and distributed? (No=high risk)	O	O	O	O
(6) Has the formulation of construction operational hours been analyzed relative to owner operation hours? (No=high risk)	O	O	O	O
(7) Determination of early morning concrete pouring operations relative to owner operations?	O	O	O	O
(8) Development of contingency plan for interruption of existing life safety systems during construction?	O	O	O	O
(9) Are off peak traffic times required for construction material deliveries? (Yes=high risk)	O	O	O	O
(10) Determination of worker and end user parking areas during construction.	O	O	O	O

Figure 35:   Risk Matrix Section 5.

5.1. Distribution of the commissioning plan to the contractor.

5.2. Distribution of the activation plan.

5.3. Distribution of equipment plans by the owner's agent.

5.4. Distribution of IT plan prepared by the owner's IT department and consultant.

5.5. Distribution of temporary wayfinding and signage plan.

5.6. Formulation of construction operational hours relative to owner's operations.

5.7. Determination of early morning concrete pouring operations relative to owner's operations.

5.8. Plan development for interruption of existing life safety systems during construction.

5.9. Material loading/unloading hours and locations during construction, and

5.10. Determination of worker and end user parking areas during construction.

These items are aimed at assigning risk levels based upon the stage of each of the plans identified above. While some of these items seem to overlap the content in Section 4, this phase of evaluation assesses the risk level of each of the micro plans. Under optimum conditions, these plans should be prepared during preconstruction and distributed to the project execution team. Once these, microplans have been distributed, the risk level can be estimated. At a minimum, high risk topics should then be addressed in the baseline schedule.

The commonality of sections three through five are the impacts to the owner's operation both during construction and activation. These conditions can apply to an adjacent existing facility, a renovation, or a new facility. Multiple purposes to all three conditions include minimizing impact and increasing understanding of the business issues of the client.

4.2. Part II: Risk Identification − Design and Construction Perspective

Now that the key issues of interest to the owner and end user have been evaluated, the focus of the survey shifts to address the perspective of the designer and construction practitioners. The items included in sections 6−10 are more typically equated with conventional construction planning. Areas of focus include as-built conditions, authorities having jurisdiction, worker accessibility within the building, team member experience, and high-risk construction elements. The key elements of each section are discussed below.

Section 6 − As Built Conditions Which Can Impact Construction (Figure 36) − attempts to identify existing conditions, seen or unseen, which create risk for the project through the following:

6.1.−6.2. Determine if underground and overhead utilities match the as built drawings.

6. ASSESSMENT OF AS-BUILT CONDITIONS WHICH CAN IMPACT CONSTRUCTION	High Risk	Moderate Risk	Low Risk	NA
(1) Have the underground utilities been confirmed to match the as built drawings?	O	O	O	O
(2) Have overhead utility lines been confirmed to match the plans?	O	O	O	O
(3) Do overhead utility lines need to be relocated to accommodate new construction?	O	O	O	O
(4) Do underground utility lines need to be capped or removed?	O	O	O	O
(5) Is sufficient drainage for storm water available during construction?	O	O	O	O
(6) Are existing overhead utilities close to the vertical airspace of the new building? (Yes=high risk)	O	O	O	O
(7) Determine if pre-existing foundations exist within the construction footprint of the new project. (Yes=high risk)	O	O	O	O

Figure 36:   Risk Matrix Section 6.

6.3. Determine if there are overhead utilities which need to be relocated prior to construction commencement.

6.4. Determine if there are underground utilities which need to be capped or removed, and is a public entity required for removal.

6.5. Determine if sufficient drainage is available during construction operations.

6.6. Identify aerial or ground conditions that could restrict material delivery to the site, and

6.7. Determine if pre-existing foundations exist within the construction foot print of the new project.

The importance of verifying hidden conditions early in the project cannot be overstated. This proactive approach can actually create positive contingency in a schedule. Specifically, the earlier the identification of conditions which do not match the planned construction creates time to mitigate the discrepancy. Early identification of mismatched conditions is a best practice that should be implemented on all construction projects. Additionally, any item which is in the purview of another vendor to perform and is not contractually bound to the contractor should signal a potential for time loss. Historical experience has shown that the relocation of utilities, either overhead or underground, by major grid providers creates pinch points in the schedule. This is due in part because these companies have a much larger priority to service a large customer base, not one construction project. Additionally, planning within the construction schedule is apt to include insufficient durations for complex or externally controlled sequences. Hence, another reason why communication between interdependent providers is critical to the development of a successful schedule.

Section 7 – Authorities Having Jurisdiction (AHJ) Risk Evaluation (Figure 37) – attempts to identify nonapparent activities involved in the acquisition of permits so that a project may begin construction. This section is particularly important because construction is typically governed by local codes so that the buildings can withstand geophysical conditions in the area of the project. While most permit granting authorities look for similar information, there are idiosyncrasies in the process for approval and the building elements required. In

7. RISK EVALUATION OF AUTHORITIES HAVING JURISDICTION	High Risk	Moderate Risk	Low Risk	NA
(1) Has the governing authority for permitting been notified of the project? (No = High Risk)	O	O	O	O
(2) Has the electrical power company been notified of project requirements for both temporary and permanent service?	O	O	O	O
(3) Are there ancillary permit services required (i.e. FAA, State, Dept. of Health, EPA)? (Yes – High Risk)	O	O	O	O
(4) Have arrangements for pre-review of the project been made with the governing agency prior to submission for permit? (No = High Risk)	O	O	O	O
(5) Determination of phased permitting packages for submission (i.e. Site, foundation, shell, and build out).	O	O	O	O
(6) Determination of the review durations and phases by the permitting agency once the plans are submitted for permit. (No = High Risk)	O	O	O	O
(7) Understand if extensive changes are required for submission after permit receipt (i.e. plan changes) (Yes = High Risk)	O	O	O	O

Figure 37:   Risk Matrix Section 7.

effect, this deregulated approach toward permitting creates a significant opportunity for differences between agencies. A simplistic example is the higher wind ratings for buildings in hurricane prone areas, such as Florida. Construction methods will differ to create a building to withstand those conditions which may not be applicable to a calmer wind environment. The following section attempt to reveal the processes and elements that the local governing authority utilizes before a project can commence.

7.1. Arrangement for pre-review of the project with the governing agency prior to submission for permit.

7.2. Notification to power company for both temporary and permanent electrical service to the project.

7.3. Identification of nonprimary permitting agencies, such as FAA, EPA, Dept. of Health, et.al.

7.4. and 7.6. Determination of the review durations and phases by the permitting agency once the plans are submitted for permit.

7.5. Determination of phased permitting packages for submission, i.e., site, foundation, shell, and build out.

7.7. Understanding if extensive changes are required for submission after permit receipt. In other words, will the design period extend into the permit phase where the drawings will be required to be resubmitted? This condition is often encountered when there is an accelerated effort to start construction.

Significant momentum can be lost in the beginnings stages of a project without attention to detail as to what must be completed to commence construction. This section attempts to bring the risks involved in preconstruction start-up to light so that adequate planning can be implemented to start the project with positive momentum. Momentum is an intangible reality of any project and must be monitored closely for the purpose of preventing productivity loss.

8. RISK EVALUATION OF BUILDING CONFIGURATION VS. WORKER ACCESSABILITY	High Risk	Moderate Risk	Low Risk	NA
(1) Determine adequate number of entrance/exits for worker and material accessibility during construction. (No = High Risk)	○	○	○	○
(2) Determine the logistical aspects by floor, based upon floor square footage area. (No = High Risk)	○	○	○	○
(3) Determine if materials can be staged in close proximity to the construction installation area. (No = High Risk)	○	○	○	○
(4) Determine if worker break areas are close to the actual work area. (No = High Risk)	○	○	○	○

Figure 38:   Risk Matrix Section 8.

Section 8 – Building configuration vs Worker Accessibility (Figure 38) – attempts to reveal any nonproductive work environments which could hamper construction efforts during project execution.

8.1. Determine adequate number of entrance/exits for worker and material accessibility during construction.
8.2. Determine the logistical aspects by floor, based upon floor square footage area. (Less than 25KSF, greater than 50KSF, greater than 65KSF)
8.3. Determine if materials can be staged in close proximity to the construction installation area.
8.4. Determine if worker break areas are close to the actual work area.

Productivity is the risk element that is addressed by the survey items above. Should productivity be hampered by not having sufficient access, then workers will be commencing their activities later than planned. The size of the space also determines how many trades can work on the same floor. The smaller the floor, the risk increases that multiple trades will impede each other's progress. This can greatly diminish any time gains based upon conflicts between the trades. Lastly, the proximity of materials to be installed and worker break areas to the actual work area can be a silent killer when it comes to time loss in a schedule. Should any of these nonpreferable conditions exist then contingency time needs to be included in a schedule as productivity will be lessened. A counter measure could be the installation of a crew who is responsible for stocking the work areas appropriately so that the work force can stay focused on the tasks within the project schedule.

Section 9 – Team Composition – Experience Levels of Team Members (Figure 39) – attempts to determine areas where the team could strengthen their ability to perform on a specific project type.

9.1.–9.6. Questions 1–6 identify if each of the team practitioners have experience in the project's type of experience. If experience is lacking, then a higher risk is present for that team member.
9.7. Determination of previous working experience between practitioners.
9.8. Identification of the basis for decision-making by project leadership.
9.9. Identification of team protocols and reporting.

9. RISK ASSESSMENT OF TEAM MEMBER EXPERIENCE LEVELS - TEAM CARICATURE	High Risk	Moderate Risk	Low Risk	NA
(1) Owners' Project Manager experience in project's type of construction	O	O	O	O
(2) Architects' Project Manager experience in project's type of construction	O	O	O	O
(3) Contractors' Project Manager experience in project's type of construction	O	O	O	O
(4) Major Subcontractors Project Manager experience in project's type of construction	O	O	O	O
(5) Engineers' Project Manager experience in project's type of construction	O	O	O	O
(6) End User's experience in project's type of construction	O	O	O	O
(7) Has this team of practitioners worked together previously?	O	O	O	O
(8) Has a basis for decision-making been established for the team by project leadership?	O	O	O	O
(9) Have team communication protocols and reporting been established?	O	O	O	O
(10) Has a kick off meeting been conducted which outlines the working parameters for the team?	O	O	O	O
(11) Is personnel staffing sufficient to meet the baseline schedule?	O	O	O	O

Figure 39: Risk Matrix Section 9.

9.10. Execution of team work parameters through a project kick off meeting.
9.11. Analysis of staffing needs vs current staff availability to ensure ability to meet the baseline schedule.

Most experienced practitioners agree that effective team composition generally equates to project success. Section 4 dealt with many of the intangibles that generally equate to team success, while this section deals with that tangible element of work experience. History has shown that it is not an absolute that team members must have project type experience that matches the building to be constructed. There have been many cases where the author has seen project teams that were new to a specific building type (i.e., hospital, warehouse, office et al.) succeed in their execution. Relative project experience does minimize the learning curve for the idiosyncrasies of that type of building. The time created from not having to learn these unique elements allows the team member to focus on problem resolution and proactive planning. This redirection of resource time can result in higher percentages of success. The same premise holds true for teams which have worked together previously. Communication patterns, capabilities to perform, and confidence levels are a known quantity at the start of a project with team members who have worked together. This inner knowledge allows speed to deal with project specific issues which in turn creates productivity. Lastly, every project staffing plan should be evaluated to ensure that the baseline schedule can be met with the most qualified personnel. Staffing has and will continue to be the most fluid resource in project execution.

Section 10 – Identification of High Risk Project Aspects (Figure 40) – attempts to identify unique elements of the project that may present risk during construction. Generally, the risk is generated from the uniqueness of the assembly, difficulty in execution, or limited availability of materials. An example from Case Study #1 was the difficulty and uniqueness of the tiara on top of the 25-story building that over hung the structure by as much as 30 feet. This condition was further compounded by the window washing system affixed to the tiara

10. IDENTIFICATION OF HIGH RISK PROJECT ELEMENTS	High Risk	Moderate Risk	Low Risk	NA
(1) Are there any materials to procure out of state or country?	O	O	O	O
(2) Have difficult design/construction components been identified?	O	O	O	O
(3) Has a sub team been designated to solve constructability issues associated with items identified in item 2?	O	O	O	O
(4) Are their sole source suppliers or installers specified? (Yes = High Risk)	O	O	O	O
(5) Are their areas of work that require extensive scaffolding to construct? (Yes = High Risk)	O	O	O	O
(6) Are their areas of work that are considered small and confined spaces? (Yes = High Risk)	O	O	O	O

Figure 40:   Risk Matrix Section 10.

structure. A specialized track system for the maintenance buggies was required due to the unusual shape of the tiara structure. The design and constructability of this major building component extended for a duration of approximately nine months. Since the component was elevated 25 stories above street level, it was critical that planning occurred long prior to installation.

10.1. Another point of risk can be the procurement of materials, either from a sole source provider or via long distance. Long distance transit times can pose a risk due to weather conditions during shipping or the additional time for delivery when compared to locally procured materials.

10.2.–10.3. Designation of a subset of the execution team charged with the responsibility of solving constructability issues. Oftentimes, the immediate need of the project is focused on present day execution and resources are not available for advance constructability planning. Insufficient resources for advanced planning will eventually create delays in the schedule.

10.4. The risk associated with single source requisition is based upon the inability to get the materials should the vendor experience delays or default.

10.5. The last point of concern is ensuring that appropriate preparation measures are accounted for in the schedule prior to actual construction. For example, scaffolding is required for the workers building a 5-story atrium space. If the atrium space is large, the scaffolding construction could take a long time to prepare relative to the overall time to construct.

10.6. Another example would be the construction of the stairways when they provide the only entrances into the building. The phasing of these spaces becomes critical when obstruction to worker access to the construction zone. Constructability studies of the confined areas, worker access, and preparation methods are important methods to avoid time loss in the schedule.

4.3.   Part III: Risk Identification – Schedule Elements

At this point in the process, the perspective of the owner, designer, and contractor have been collected in terms of risk to project execution. Now, it is time to transcribe this data into identifiable activities within the schedule, so the focus on risk management will be constant through project completion. The sections below are presented as WBS components within the schedule and include generic risk specific parameters. In some respects, sections 11–21 can be used as checklists to confirm that a project schedule contains activities which relate to each parameter/definition.

Section 11 – Contractual Requirements WBS (Figure 41) – attempts to identify and track milestones directly related to the contract between the owner and general contractor. During 38 years of experience, few schedules have included this key information and yet, it is the most significant data which needs to be tracked in order to determine schedule success. Practice during the last 5 years as both a scheduling consultant and general contractor have shown the most effective location for these activities is in the very beginning of the schedule. Since most schedule reviews commence on the first page, this reinforces the importance of the contractual milestones and facilitates ease in comparison to actual progress. The key elements involve identification of.

11.1. Activities which require owner approval.
11.2. Black-out periods of work as mandated by the owner to preserve onsite operations.
11.3. Inclusion of the projected substantial completion date based upon construction updates to the schedule.
11.4. Substantial completion date based upon original contract duration for construction.
11.5. Owner-approved weather days of delay.
11.6. Contractor requested weather days of delay.
11.7. Updated substantial completion based upon original contract duration, approved weather delays, and approved additional scope duration.
11.8. Interim phasing as agreed upon between owner and contractor.
11.9. Owner-approved duration for additional scope of work, and

11. CONTRACTUAL REQUIREMENTS - WBS	Not Defined	Poorly Defined	Well Defined	NA
(1) Identification of all owner approval items per the contract.	O	O	O	O
(2) Identification of blackout work periods by the owner, in order to not impact operations.	O	O	O	O
(3) Identification of projected substantial completion.	O	O	O	O
(4) Identification of contractual substantial completion.	O	O	O	O
(5) Identification of actual delays to the schedule (during construction).	O	O	O	O
(6) Identification of requested delays to the schedule (during construction).	O	O	O	O
(7) Identification of contractual duration of the project per the contract.	O	O	O	O
(8) Identification of any interim completion date requirements (i.e. phasing)	O	O	O	O
(9) Identification of additional scope contractually added to the project (during construction).	O	O	O	O

Figure 41: Risk Matrix Section 11.

It is recommended to include a milestone for projected substantial completion that is linked to the actual contractor-controlled work. Inclusion of this projection allows the reviewer to easily assess progress vs contractual requirement. Feedback has indicated owner's, architects, and contractor find this beneficial because it quickly focuses the conversation on assessing schedule status. Speed in realizing the situation also promotes faster problem-solving, which can minimize risk.

Section 12 – Design Requirements WBS (Figure 42) – attempts to identify key design deliverables which can affect construction. The design elements involve identification of:

12.1. Confirmation of room numbering protocol.
12.2.–12.3. Design package owner review, approval periods, and plan adjustment time for changes,
12.4. Enumeration of phasing design packages.
12.5. Assessment of design staffing to meet the baseline schedule, and

Intuitively design must be completed prior to construction commencement however, some of the deliverables often occur after groundbreaking (room numbering). Room numbering operations can have an impact on MEPF BIM design nomenclature and label identification of installed systems. As BIM coordination generally occurs early in the project, it is important that this deliverable is defined concurrently. Other questions relate to the capacity of the design team to meet the baseline schedule due dates. As discussed in the Case Study #1, design staffing can have a huge impact on the ability to meet execution schedules. The construction procurement strategy is also important relative to phasing design packages. These incremental target dates also impact staffing management and construction commencement. Lastly, owner approval time is essential to understand and accommodate in the baseline schedule. The speed of the institution or owner's agency is oftentimes unaccounted for in the baseline schedule.

12. DESIGN REQUIREMENTS - WBS	Not Defined	Poorly Defined	Well Defined	NA
(1) Has a room numbering protocol been defined for the purpose of identification of MEPF systems?	O	O	O	O
(2) Identification of owner approvals during design document production.	O	O	O	O
(3) Inclusion of adjustment periods required to update design documents with owner comments.	O	O	O	O
(4) Identification of phased design packages (if applicable). Insure that scope is aligned within packages to ensure constructability with segmented permits.	O	O	O	O
(5) Assessment of sufficient staffing to meet phased design packages.	O	O	O	O

Figure 42:   Risk Matrix Section 12.

13. PERMITTING REQUIREMENTS - WBS	Not Defined	Poorly Defined	Well Defined	NA
(1) Allowance in the schedule for at least two submissions to the AHJ prior to receipt of permit to commence construction.	O	O	O	O
(2) Allowance for AHJ comments/revisions to plans between submissions.	O	O	O	O
(3) Analysis of capacity of AHJ Resources compared to anticipated review period by AHJ.	O	O	O	O
(4) Review of other materials required for submission by AHJ with each package (i.e. energy code compliance).	O	O	O	O

Figure 43:   Risk Matrix Section 13.

Section 13 — Permitting Requirements WBS (Figure 43) — attempts to identify the subtle time losses incurred during the permit process. The focal points are:

13.1. Anticipation of two review periods by the AHJ.
13.2. Identification of time periods for plan modifications between review periods by AHJ.
13.3. Identification of review period anticipated by AHJ vs. anticipated work load by AHJ, and
13.4. Identification of other materials (other than plans) which are required for review by AHJ.

The risk associated with permitting operations are generally high, since completion is beyond the control of the design and construction practitioners. This condition places a priority for preplanning communication between designers and AHJ permit officials. Prudent management includes conferencing with permit reviewers ahead of permit submission for the purpose of answering as many questions as possible. Face-to-face interaction also creates engagement by all parties which increases the probability of expeditious permit processing. This area of the schedule is often underestimated by contractors, since pre-submission research is seldom implemented. Since work cannot start without an approved permit, the more knowledge obtained about this process will generally result in a more realistic schedule.

Section 14 — Owner Procurement Requirements WBS (Figure 44) — attempts to illuminate all scope of work that is contracted directly by the project owner. This section shares similarity with the permitting in that the work is beyond the control of the contractor. A major difference between Sections 12 and 13 is that owner procurement generally involves coordination by the contractor in providing infrastructure connectivity to owner procured items. Most contractual requirements state that time must be included in the schedule for coordination and installation of owner provided scope. To assess the appropriate time for this type of work, this section focuses on the following elements:

14.1. Identification of owner vendors shop drawings, approvals, manufacturing, delivery, and contractor supplied infrastructure.

14. OWNER PROCUREMENT REQUIREMENTS - WBS	Not Defined	Poorly Defined	Well Defined	NA
(1) Identification of pre-purchased equipment by owner which requires construction installation by the contractor.	O	O	O	O
(2) Identification of owner supplied utility services.	O	O	O	O
(3) Identification of owner furnished and owner installed scope of work. (OFOI)	O	O	O	O
(4) Identification of owner furnished and contractor installed scope of work (OFCI).	O	O	O	O
(5) Identification of External approvals by owner contracted vendors.	O	O	O	O
(6) Identification of legal requirements prior to construction commencement (i.e. easements or land plats).	O	O	O	O
(7) Identification of travel time for out of state equipment service and installation.	O	O	O	O

Figure 44:   Risk Matrix Section 14.

14.2. Identification of owner's provision of utility services with contracted providers.

14.3. Identification of timeline for owner supplied and installed scope.

14.4. Identification of timeline for owner supplied and contractor installed scope.

14.5. External approval by owner vendors of installation drawings which coordinate infrastructure of both the contractor and vendor.

14.6. Identification of owner responsible legal requirements in terms of easements and right of ways for utility services, and

14.7. Inclusion of travel time from long distance suppliers.

Experience with platting property and easement designations reveals another layer of participants with the legal profession. This process can be particularly cumbersome and time consuming. As these easements are usually critical in providing space for utility services, a project with this requirement can be at risk if not handled in a timely fashion. The theme of work execution sans contractor contractual engagement places particular concern for time slippage.

Section 15 – Bid/Pricing Requirements WBS (Figure 45) – enumerates the steps by the contractor to finalize the contract prior to construction commencement. The particular items listed in section fifteen below assume the construction procurement method is Construction Management at Risk (CM@R). In this delivery model, the contractor acquires subcontractor bids and then submits a Guaranteed Maximum Price not to exceed. Upon submission, the owner and design team review the estimate and if it is over the budget, the value engineering (VE) process commences. This process involves the submission of cost savings for equal quality materials. Once the list of VE is priced and reviewed, the budget and cost can be realigned to meet expectations and start construction. The notice to proceed (NTP) can be issued anytime during the process to facilitate timely procurement of materials so the schedule dates can be accomplished. These steps are the last crucial activities that must be completed prior to construction commencement. Depending upon the size of the project, this process can span two to six weeks. Oftentimes, contractors only account for their time in preparing the bids. In fact, the review and acceptance period can be long

15. BID/PRICING REQUIREMENTS - WBS	Not Defined	Poorly Defined	Well Defined	NA
(1) Identification of bid period to subcontractors, assembly time period, and report preparation.	O	O	O	O
(2) Identification of owner bid review and approval.	O	O	O	O
(3) Allowance for cost reduction/value engineering period post bid review.	O	O	O	O
(4) Issuance of Notice to Proceed (NTP).	O	O	O	O

Figure 45: Risk Matrix Section 15.

16. CONTRACTOR PROCUREMENT REQUIREMENTS - WBS	Not Defined	Poorly Defined	Well Defined	NA
(1) Inclusion of mock up preparation and approval periods.	O	O	O	O
(2) Inclusion of shop drawing production, approval, manufacturing, and delivery to site.	O	O	O	O
(3) Inclusion of BIM coordination.	O	O	O	O

Figure 46: Risk Matrix Section 16.

durations especially if the bids do not fall within the owner's budget. It is critical that these steps are identified:

15.1. Duration of bid period by subcontractors, contractors time to assemble the total estimate, and create a reviewable report.
15.2. Duration of owner bid review and approval.
15.3. Duration of potential VE period, post bid review, and
15.4. Issuance of NTP.

Section 16 – Contractor Procurement Requirements WBS (Figure 46) – identifies the coordination processes associated with Building Information Modeling (BIM), Mock-Ups, and shop drawing/submittals associated with critical path activities in the construction schedule. The commonality between these items is that they must all be procured and coordinated before actual construction using these identified building materials can commence. The intent of this section is to identify high risk procurement activities as follows:

16.1. Physical mock-ups and approvals prior to procurement.
16.2. Supplier shop drawing preparation, approvals, manufacturing, and delivery to the jobsite, and
16.3. BIM processes and production on MEPF systems.

The emphasis on this section is to identify construction materials of high risk and their comprehensive procurement operations and pre-installation coordination. For example: mock-up preparation is the coordination of several building materials in their final conditions to ensure that they function according to specification. An exterior wall mock-up is appropriate when there are several waterproofing materials that come together at difficult angles. The intent is to test their ability to prevent water from penetrating the prescribed materials in their final configuration. The leak test of dissimilar materials which replicates their intended installation on a building allows contingency time to test and develop the installation to ensure weatherproof performance. This contingency time can then be used to modify the design should the waterproofing aspect of the mock up not perform as planned.

The entire process for shop drawings is identified which includes production, approval, manufacturing, and delivery to the jobsite. Typically, the contractor tracks due date for shop drawings in the schedule leaving the other time-consuming elements to chance. It is recommended practice to track all of the steps for high risk material procurement activities. During the review of Section 16's matrix, it is important to identify those high-risk procurement materials, so the process can be included in the baseline schedule.

Lastly BIM coordination for the structural, MEPF protection systems is important to streamline installation. This modeling process must be completed prior to actual construction installations to be effective. Effectiveness is defined as providing proper clearances between all system pipes and ducts so that all elements have room within the chases. By modeling the location of each of the systems relative to their location in the building space, efficiencies can be gained by running piping more efficiently with less waste. Another benefit is to minimize crash detection, where dissimilar systems may end up trying to be located in the same space. The model approach creates a more interactive and proactive means to solving material locations which don't collide with dissimilar systems. All of these processes have potential to create schedule slippage if the multiple aspects of the process are not adequately documented and monitored.

Section 17 – Construction Activities Requirements WBS (Figure 47) – attempts to identify key milestone dates during the typical building process. The significance of each milestone is that it typically acts as a commencement point for subsequent construction sequences. For example, temporary dry-in must be complete before build-out can commence without impact from weather. The common milestone progression points are itemized below with their successor construction activities:

17. CONSTRUCTION ACTIVITIES REQUIREMENTS - WBS	Not Defined	Poorly Defined	Well Defined	NA
(1) Agreement on weather days allowed and documentation of actual weather days in the schedule.	O	O	O	O
(2) Identification of temporary Dry-in date.	O	O	O	O
(3) Identification of permanent Dry-in date.	O	O	O	O
(4) Identification of temporary air supply for interior finishes.	O	O	O	O
(5) Identification of conditioned air vs controlled air date.	O	O	O	O
(6) Identification of permanent power date.	O	O	O	O
(7) Identification of electrical and HVAC systems go live date.	O	O	O	O
(8) Identification of Floor moisture rating achievement date.	O	O	O	O
(9) Inclusion of pretest operations (i.e., pier depth).	O	O	O	O

Figure 47:   Risk Matrix Section 17.

17.1. Inclusion of actual weather days in the contractual area of the schedule.

17.2. Identification of temporary dry-in date so that nonmoisture impacted activities in the build out phase can commence.

17.3. Permanent dry-in date and conditioned air milestones with successors of millwork, acoustical ceilings, and paint.

17.4. Temporary air supply to allow installation of drywall systems.

17.5. Controlled air which allows commencement of technical air balance commissioning.

17.6. Permanent power by electrical company which precedes.

17.7. Electrical and HVAC systems going live.

17.8. Moisture testing of concrete floors to confirm ability to apply tile finishes, and

17.9. Pretesting required for piers or soil to confirm appropriateness of design prior to construction.

Lastly, it is critical to come to an understanding of the weather days allowed per reporting period, if not identified in the contract. All construction-related activities would be included within this WBS of the project schedule. To ensure a reliable schedule, these milestones should be linked with the appropriate logic relationships.

Section 18 – Commissioning Requirements WBS (Figure 48) – focuses on contractor activities that are related to final commissioning by the owner's agent. An increasing trend in contractual scheduling contract requirements is the provision of including commissioning as a requirement for substantial completion. Owner's typically employ an independent agent to perform the commissioning services, so this applies a high element of risk since production of this task is not directly controlled by the contractor. Additionally, commissioning can vary greatly from project to project based upon the complexity of the building systems. The level of detail will also vary as some clients require only spot checks of the various systems, while others require all components of the systems to be verified. Lastly, since this process occurs during the final stages of a project,

18. COMMISSIONING REQUIREMENTS - WBS	High Risk	Moderate Risk	Low Risk	NA
(1) Is the commissioning agent under contract with the owner during the design phase? And final commissioning report required for substantial completion?(Yes = High Risk)	O	O	O	O
(2) Is chemical flushing of water systems required?	O	O	O	O
(3) Is commissioning required for 100% of all systems? (Yes = High risk)	O	O	O	O
(4) Is integrated systems testing required?	O	O	O	O
(5) Are control systems designed to meet specific owner needs?	O	O	O	O
(6) Is the Ethernet system supplied by the owner (required for control systems)?	O	O	O	O
(7) Is leak testing required for air and water systems?	O	O	O	O
(8) Are their requirements for special systems testing (i.e.,ammonia, pressure, hydro)?	O	O	O	O

Figure 48: Risk Matrix Section 18.

there is a tendency to wait until the project is well under construction before addressing commissioning planning. All of these factors create risk which increases the importance of planning proactively for this process. Below are some of the considerations to be addressed in the commissioning section of the schedule.

18.1. Determine when the third-party commissioning agent will be integrated into the planning process. Proactive planning opportunities can be missed if the agent is not involved during the design process. Both cost and time can be impacted if discussion about the systems commissioning process is not addressed during the design phase.
   Determine if the owner requires the systems commissioned as a requirement for obtaining substantial completion. Time requirements can be impacted greatly during commissioning since most of the operations is a test, adjust, and retest operation. Schedule variability can span a broad range based upon retesting and the complexity of the systems being tested.

18.2. Identify if chemical flushing is required for mechanical water systems. Completion of this chemical interaction process can vary depending upon the existing water systems and the required water composition readings.

18.3. Determine the extent of commissioning and testing on all building systems. Are all components required to be tested (100%) or is spot testing acceptable (10%). This factor can vary greatly between clients and based upon the complexity of the systems. One of the most common elements of delays is lack of clarity of this requirement before subcontractors are engaged in the project. Both time and money can be significantly impacted if this requirement is not understood early in the project.

18.4. Integrated systems testing is the ability of multiple systems to work together as designed. For example, a control system may want an air vent to initiate after a pump has commenced operation. In this example, a mechanical system would interface with a plumbing system through the low voltage controls. Obviously, this involves more subcontractors to work together with the designers to ensure that they systems will respond according to the owner's desired operation. A general rule of thumb in scheduling is that as the people count increases so do the scheduling risks. Therefore, this should be a trigger to allow for additional planning so that a reasonably accurate schedule can be developed.

18.5. A final discussion on how the control systems are designed to meet the owner's needs should be conducted including the contractor. Oftentimes, this is taken for granted since the team assumes that it was taken care of during the design phase. This discussion serves a twofold purpose of bringing the contractor up to speed with the rest of the team and integrating any evolutionary changes into the design. Control systems are the newest frontier for the building industry and as with other aspects of technology, changes occur at a faster rate than most construction methods. The authors experience in this area is that changes in expectations happen more times than not. The loss of time in making corrections while executing the project can be a large risk that can be avoided if addressed early in the project.

18.6. Verification of procurement authority for the building control Ethernet system is important especially if it is not under the direct control of the contractor. If the owner is integrating the new project into an existing campus, then the tendency is for the owner to procure the system to ensure compatibility. Common thinking indicates additional time and planning are necessary in the schedule for items not under the specific control of the building contractor.

18.7. Determine if a formal leak test is required for water and air systems.

18.8. Lastly, the commissioning of sub systems, i.e. ammonia, pressure, hydrotesting, should be detailed within the schedule. Items which are considered special systems bring an element of uniqueness which can present unforeseen issues during execution.

Compared to previous risk elements in this book, commissioning poses the greatest potential for delay based upon the additional parties not directly under the contractor's purview. Contractual terms with regard to commissioning requirements are critical to understand prior to baseline schedule development. Clarity, collaboration, and cooperation are instrumental to successfully meeting schedule goals.

Section 19 – Inspection Requirements WBS (Figure 49) – attempts to identify AHJ and associated final acceptance of construction installations. The construction industry is structured where AHJ's are decentralized to local levels so that specific area building parameters can be enforced. The lack of centralization, by nature, creates multiple variables of compliance. These variables are all risk opportunities which need to be accounted for in the project schedule. A best practice involves meeting with the AHJ prior to groundbreaking to determine the requirements for construction approvals. The items in Section 19 are intended to be potential questions to the AHJ.

19.1. Identification of owner inspections is relative when the owner is also the AHJ. An example of this condition is when the owner is a state agency.

19. INSPECTION REQUIREMENTS - WBS	Not Defined	Poorly Defined	Well Defined	NA
(1) Identification of Owner inspections.	O	O	O	O
(2) Identification of AHJ inspections to achieve owner occupancy.	O	O	O	O
(3) Identification of owner shut downs/tie ins.	O	O	O	O
(4) Identification of AHJ cover-up inspections to continue construction work.	O	O	O	O
(5) Identification of final inspections to achieve life safety protection of facility.	O	O	O	O
(6) Identification of conditions/inspections required to achieve substantial completion. (i.e., punchlist preparation)	O	O	O	O

Figure 49: Risk Matrix Section 19.

Typically, the state agency is exempt from city jurisdiction inspections and the state agency will provide their internal inspections.

19.2. Understanding the specific inspections required for owner occupancy is critical to obtaining substantial completion. The industry definition of substantial completion is the time in which the owner can occupy the building for the intended use. This trigger date can have financial deficits to the contractor if the deadline is not met. Definition of these key finish milestones in the schedule and application of proper logic sequencing is paramount for a successful project.

19.3. Owner shut downs/tie-ins relate to utility connection activities between an existing and new system. These activities can be schedule killers if not adequately planned. Research as to what the existing service line impacts is required as well as defining an appropriate time in which the service can be disrupted. End users who benefit from the service will take priority in their usage over the needs of the contractor. Hence, careful planning is a must to reduce time loss.

19.4. Cover-up inspections are those intermediary inspections required during the construction process and are not typically identified in the schedule. An example would be inspection of in wall utility systems prior to sheetrock cover up. When preparing the schedule an assessment of the quantity and duration of these inspections should be made to determine if they should be noted in the schedule. The larger the number of inspections would indicate the need for inclusion in the schedule. It is common to include this time in the actual activity of the item being inspected. Risk management would suggest that these items be identified separately if the number of inspections are voluminous and dedicated schedulers are available to update the schedule.

19.5. Life safety inspections are defined as those that ensure occupants safety within the building. Without exception, substantial completions are granted only when all life safety elements are inspected and approved. While this is only one aspect of achieving substantial completion, it is considerable and should be treated with due care. This can include fire sprinkler systems, emergency control systems, backup generators, and fire doors. Examination of the particular components within each project should be diligently identified and provided for as activities in the schedule.

19.6. Substantial completion may seem redundant at this point, however, nonbuilding elements are associated with achieving this condition. Specifically, architects require that the punchlist be agreed upon by the owner, contractor, and themselves prior to achieving substantial completion. Each contract should be fully understood as to what requirements are necessary to achieve substantial completion and then identified within the schedule.

Inspection processes can become nonproductive to the schedule if they are not planned for or if reinspections are required for the same work. It is the contractors job to understand all processes required especially those which are beyond

20. ACTIVATION REQUIREMENTS - WBS	Not Defined	Poorly Defined	Well Defined	NA
(1) Process identification for end user notification of incorrect installation and tracking system for corrections.	O	O	O	O
(2) Provision for vertical transport of FF&E installation in conjunction with punchlist completion.	O	O	O	O
(3) Provision for down time by operations during relocation of existing equipment.	O	O	O	O
(4) Evaluation of move in duration by owner activation committee.	O	O	O	O
(5) Move in action plan completion by activation committee.	O	O	O	O

Figure 50:   Risk Matrix Section 20.

their direct control. Time allocated for tracking these activities is well spent and can avoid delays at the end of the project.

Section 20 – Activation Requirements WBS (Figure 50) – attempts to identify owner activities related to occupancy of the project. At first glance, most professionals would make the assumption that this does not impact the construction schedule. However, in the event of phased openings of a project, these actions by the owner can impact the construction schedule. In particular, if one area must be finished and the owner must move in before the second area can be constructed. This would most typically occur in hybrid projects where new and remodeled construction are involved in the scope.

20.1. One of the most difficult aspects of new project occupancy is adaptation to the space by the end users. This process can be difficult on the owner's facility personnel especially if the company has not experienced a move process with the staff on board. Based upon a move assessment, it is important to have procedures in place which allow for an easier transition.

20.2. If the project is a multistory building, then vertical transportation during the activation period is an important consideration. When you consider that the highest volume of people working in the building is during the activation phase, mobility is essential for movers, end users, and construction workers. A vertical transportation plan is advisable prior to building impact so that construction completion and move in operations are not adversely impacted.

20.3. Major equipment relocated from existing facilities creates operational downtime with current operations. Additionally, it may take longer to breakdown, transport, and reinstall the equipment due to the nonoptimal packaging. Typically, the owner will require that the out of service duration is minimized to reduce financial impacts. This timeline situation generally requires that this type of equipment be transferred at the end of the construction process. Based upon these factors there are more variables associated with this process that can create schedule risk.

20.4.–20.5. Move duration is generally managed by the owner's staff and consultants which means that it is not within the control of the contractor. While the contractor's primary milestone is substantial completion, a satisfied client is created by meeting their intended use date for services rendered. That condition heightens the importance of planning the move/activation period. In fact, the author has experienced the condition where substantial completion was late, but advanced planning shortened the activation period such that the project finished on time.

Since clients remember how you finish, participation in this endeavor can serve to solidify relationships as well as achieve successful final completion. The circumstances that precipitate the need for move planning are not as important as facilitating a successful ending of the project for the client's business services. The author has experienced a range of contractors who perceived this step as important as well as those who do not share that opinion. Invariably, the contractor who collaborates with the owner to achieve a successful business opening date is more likely to achieve repeat business with the owner. If this is not sufficient motivation to perform this planning, then at the least, it should be incorporated during a hybrid remodel/new construction project is the model for execution.

Section 21 – Closeout Requirements WBS (Figure 51) – attempts to provide the necessary steps for final completion of the project per contractual obligations.

21.1. Final completion is the trigger date for releasing final funds to the contractor so the importance does not need to be emphasized. Again, each contract may contain different elements so review of the requirements prior to construction commencement is recommended. Typically, punchlist completion, final as built drawings, warranties, and lien releases are required for final completion. It is common that the duration between substantial and final completion is defined in the contract.

21.2. Third-party commissioning agents may be required to submit final reporting prior to contractual closeout of the project. This can be problematic for the contractor since there is no contractual relationship between them and the commissioning agent. The author has experienced projects where final monies were withheld pending the receipt of the final commissioning

21. CLOSEOUT REQUIREMENTS - WBS	Not Defined	Poorly Defined	Well Defined	NA
(1) Identification of contract requirements for final closeout submission.	O	O	O	O
(2) Identification of reports required from third party agencies for final closeout.	O	O	O	O
(3) Identification of conditions required to achieve final completion (i.e., punchlist completion, utilities and insurance transfer to owner).	O	O	O	O

Figure 51:   Risk Matrix Section 21.

report. Based upon this condition it is advisable that a baseline schedule establishes expected milestone dates for delivery of these documents.

21.3. In order to meet milestone dates, it is necessary to track them. The author recommends that all deliverables be identified in the last section of the schedule to aid in production of final completion documents. These deliverables are often included in Division 1 of the specifications.

As with most endeavors, it can be a struggle to finish because the contractor is moving onto the next project. In the case where different team members from the contractor may be assigned the responsibility of closing out a project, it is imperative that expectations are easily referenced for a successful closeout. My experience has been that owners remember "how you finish, not how you start," so inclusion of this WBS section cannot be overstated.

Leveraging this Risk Matrix requires gathering information from key project execution personnel and then discourse to establish alignment of understanding. Utilizing this method fosters greater understanding amongst a diverse set of practitioners, so the goal is understood from the commencement of the project. The product of better understanding should yield a seamless and smoother execution phase of the project. Identification of higher risk elements also aids the team to focus on contingency plans. Creating a culture of risk awareness by all practitioners of the project execution team can increase the percentages of success.

Chapter 5

Communicating the Project Schedule and Change Management

To this point in the manuscript, the content and context of construction risk have been discussed in detail. One of the largest risks in any endeavor is communication between all team members and delivery partners. This chapter's focus is communication techniques to ensure that laymen and industry professionals understand construction schedule progress and the associated risks prior to impact. Proven techniques which effectively communicate schedule information and methods to minimize risk during the construction process are provided in Sections 5.1–5.6 below.

5.1. Part I: Communicating through Shared Language

Common understanding is the cornerstone of effective communication. Effective communication leads to timely decision-making. Timely decisions reduce the potential for risk impact. Prior to construction commencement, it is imperative to establish the common language for the project with all team members. This process achieves three purposes; breaking the project into smaller and more manageable parameters, agreement on how the areas of the project will be referred to during construction, and development of phasing during execution.

Figure 52 demonstrates the concept of breaking a larger project into smaller more manageable areas. In this example, the size and use of the spaces were encumbered with constant pedestrian traffic so the driving question before construction was, "Where do we start?". The planning team required three months to devise an effective construction sequence that minimized existing operations. The focus for the team was to divide the areas into the highest traffic areas versus back of house service areas. The next factor for consideration was the difficulty factor in constructing the space. Lastly, the goal was to provide quick turnarounds of the spaces so that they could be utilized by management operations. As the team drilled down through each of the priority areas, phasing became clearer, and the smaller areas began to emerge as the best means to construct the project. Collaboration with all team members created consensus which allowed each professional firm to synchronize solutions during construction. This zone map became the central reference point for all construction schedules, meeting agendas, RFI's, and owner meetings of the project.

Phasing, or the timing of execution by project area, becomes the evolutionary next step in preparing the Project Language Map. The phasing options presented via the arrows in Figure 53, show two different approaches in performing

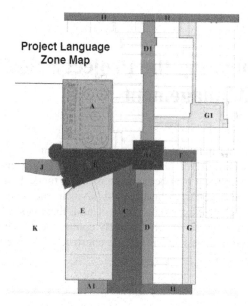

Figure 52:   Project Zone Map.

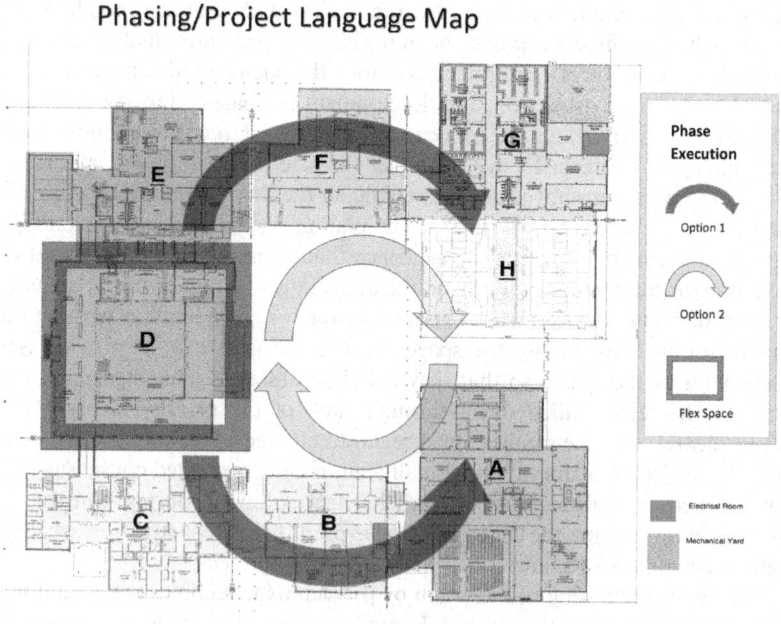

Phasing/Project Language Map

Figure 53:   Project Zone Map and Construction Flow.

the work. It is important to note that all of the buildings, except for Building D, are new construction. Building D is an existing building which will be renovated. Hence, the natural flow allows the renovated space to be designated as "Flex Space" where work crews can fluctuate depending upon availability. This strategy minimizes disruption to the construction process since work in the Building D will not be impacted by environmental conditions. Red and blue arrows provide two different approaches to construction workflow. The blue arrows indicate a clockwise rotation of construction crews which would start at Building E and complete at Building C. The approach by the blue arrow designation indicates that only one crew would be utilized to complete the entire construction scope. The red arrow process suggests the utilization of two crews where one would start at Building E and the other at Building C, working simultaneously toward Buildings, H and A, respectively. In some respects, the red arrow approach gives the execution team a contingency plan to accelerate production of the construction, if the resources are available. Regardless of which approach is chosen, the fact that a well-considered execution plan is documented prior to construction commencement can lead to greater flexibility when issues occur.

5.2.   Part II: Achieving Mutual Agreement with the Baseline Schedule

In the author's 38 years of experience, it is safe to state that at least 95% of contracts required the submission of a construction schedule for approval within 30–60 days of commencement. The American Institute of Architect's standard General Conditions document, which is typically included with the majority of commercial building projects, provides a clause which states that the contractor shall provide a construction schedule for the work (A201, Section 3.10.1). It is common for the owner to modify the general conditions to indicate the exact duration from project commencement that the original schedule should be submitted for review. This original submission, once agreed upon by the owner and contractor, becomes the reference point for all future schedule adjustments. The most common industry name for this schedule is the "baseline."

The information that is typically not included in the contract document is the instrument by which the baseline schedule is accepted. The lack of detail in formalizing the project baseline schedule can create additional delays. The author has developed guidelines for completing the process. The steps are described below:

- Contractor should develop the schedule, with participation from key stakeholders (subcontractors, vendors, support teams), during the period prior to the first pay application.
- Once the draft is developed, the contractor should review the critical path of the schedule to ensure that the logic flow is appropriate for the project. Within this review there are technical parameters which should be met, but these will be discussed in Chapter 7. Also, confirm that float is included in the

schedule per the contractual terms. Obviously, this step needs to allow time for any adjustments to the schedule based upon the review.

- Upon determination that the schedule is ready to submit as the baseline, print two schedules: one with just the critical path indicated and the second with all activities.
- Submit the two schedules to the owner with the request for approval.
- Figure 54 demonstrates the prototypical elements which should be included in the request for approval of the baseline schedule. The format of the approval is nonconsequential, so either email or letter will suffice.

During the author's tenure as a professor, many students posed the question as to whether the contractor benefited from stalling the baseline approval process. There are many reasons why the answer to the previous question is "no,"

Re: Acceptance of the Project Baseline Schedule

In accordance with the terms and conditions of the Construction Manager at Risk Agreement, Section xx, Paragraph x, "Contractor's Name" has provided the baseline construction schedule for the "Project Name" with a data date of M/DD/YYYY.

The schedule is based upon an initial start date of "xx" and a substantial completion of "xx". The scope of the schedule is for "describe scope". It is understood that the critical path of the schedule is set at zero days. The total number of work/calendar days is "xx".

(Select One Option): Schedule contingency of "xx" days is included within the activities and will be amended as the schedule progresses or a schedule contingency activity of "xx" days is included just prior to substantial completion and will be amended as the schedule progresses.

Weather days included in the schedule at the rate of "x" per day or per the contractual agreement. Weather days will only be requested as a time extension in the event they exceed the agreed upon monthly allocation included in the schedule.

After approval of the baseline schedule, the updated schedule and executive summary report will be submitted on the "xx" day of each month. "Contractor" recommends that the schedule be reviewed by the team during the first Owner, Architect, and Contractor (OAC) meeting immediately following the reports publication.

For the purpose of acknowledging acceptance of the baseline schedule as submitted, please either sign below and return or send an email confirming approval. Should written confirmation of approval not be received within 10 days, then "Contractor" will assume the baseline schedule is acceptable.

Attachments: Baseline Schedule data date

Accepted by:

_____   _____
Owner's Representative   Date

Figure 54:   Baseline Acceptance Letter.

but the primary driver is based upon the ability to defend a time extension claim through the legal process. Construction case law has repeatedly ruled that time extensions are not sustainable without the establishment of a baseline schedule through mutual agreement between the owner and contractor. Of course, this is considered the worst-case basis for a time extension to the schedule. Most projects do not devolve into legal proceedings. However, the fact remains that before any claim for time can be perfected the original baseline must be agreed upon by all parties to the contract. The author's stance on good business practice is that a team cannot effectively achieve the time goal without a clear understanding of the baseline schedule. Success is seldom achieved when team member alignment is not realized.

5.3. Part III: Periodic Schedule Reporting

The author's experience in a robust commercial construction market of the US has demonstrated that contractors have a tendency to place more effort in the preparation of the baseline schedule and less in the maintenance and communication of progress. This tendency is disturbing from multiple perspectives. As discussed in Chapter 1, change is the norm, not the exception. In order to proactively manage the evolutionary state of the project, it is imperative that a consistent mechanism is in place to monitor and adapt the schedule. Changes in workforce, building conditions, and owner constraints will happen prior to completion. The Monthly Schedule Update Report, Figures 55 and 56 are the author's chosen mechanism and was developed to meet the following goals:

- Comparison of approved baseline against progressed schedule.
- Consensus between owner and contractor on lost and delay days to date.
- Proactive reporting on potential delays so the team has the opportunity to mitigate prior to occurrence.
- Documentation of actual and proposed mitigation measures by the contractor.
- Summarization of agreed upon time extensions to the contract.
- Provide contractual documentation of requested time extensions.
- Major milestone accomplishments, and
- Most importantly, concise summarization of the updated schedule in terminology that is understood by executives and daily project team members.

This report was developed and tested over a 10-year period and has proven to provide the right amount of information to a broad range of team members. "Right" is defined as the information which provides schedule project metrics and situational content to facilitate proactive decision-making and mitigation. The challenge in communicating the schedule information is typically there are only a few team members who understand a critical path gantt schedule. For the most part, those knowledgeable team members are not the final decision-makers for the project. When the gantt critical path schedule is the only means of

Monthly Schedule Report ▌

Table1.Baseline Schedule Comparison from Previous to Current Month

% Complete	Original Baseline SC	Delay Baseline SC	Rain Days [2]	Current Substantial Completion [3]	Rev. Float[1]	Variance (CD's) From Last Month
92	11/29/16	NA	15	1/7/17	0	0

Note: 1 Revised Float reflects the cumulative float after rain days and delays are tallied for each month's schedule

Note: 2 Rain days beyond the allocated total per month to date

Note: 3 Contractual Substantial Completion based upon rain days to date

1. Contractor is currently projecting a substantial completion date of 1/7/17 consistent with the rain days and delays granted to date.
2. Contractor has documented a delay due to the impact of THIRD-PARTY MEP PROVIDER on enabling conditioned air. This delay was approved as CPR #056 and is included in table 1 above.

Table 2. Weather Days & Float Calculations based upon Weather Only

	Month	Year	WD's per Contract	WD's Actual [2]	Net WD's	Impact Critical Path	Previous Approved Float	Revised Float w Months WD's[3]
Original Rain Days Per Schedule = 44 Days	10	2016	4	0	0	NO	29	0

Note: 2 Exhibit W – Weather Day Request. Note: these float days are based upon weather usage to date and do not reflect actual slippage in the schedule for activity slippage other than for weather days unable to work.

Note: 3 Float days shown in this chart are based upon weather days approved from OWNER and exclude the actual status of float days based upon slippage or gain in any of the schedules

General Comments: Refer to schedules for key points of discussion for each project.

1. See items below regarding AT&T.
2. Items where potential scope of work is projected to add to the project are not shown on the schedule until such time that the CHANGE ORDER is approved by OWNER.

Actual Delays to the Schedule:

1. Please see table 1 for weather days incurred to date. THIRD PARTY MEP PROVIDER impact was approved as CPR #056. Revised substantial completion date including weather and THIRD-PARTY MEP PROVIDER delays is 1/7/17.

Potential Pending Delays:

3. Fully functional AT&T service was needed by 10/14/16 in order to schedule elevator final inspection as part of obtaining the certificate of occupancy. AT&T service was received on 11/10/16. Elevator inspection has been scheduled for 12/22/16 which is the earliest available. Otis elevator subcontractor is working to improve the stand-alone elevator inspection date by either getting an overtime inspection or improving normal inspection date. Contractor should know if this mitigation effort is successful prior to the thanksgiving holiday. The previous month's schedule report had a deadline of December 9th for this elevator inspection. The new projected date would represent a delay of 13 thirteen days. Additional risk associated with this issue is the fire marshal final inspection and building final inspection must occur after the elevator inspection. Should these activities incur further issues, the schedule could get pushed beyond current substantial completion date.

November 15, 2016

Figure 55:   Monthly Schedule Report Page 1.

Monthly Schedule Report

Mitigated Delay Accomplishments:

4. Not applicable this month.

Proposed Methods to Mitigate Delays to the Schedule:

1. Dental Chair Delay: Contractor has discussed the situation with the COH Field Inspectors with regard to the existence of infrastructure for the dental chair preventing a TCO. The Field inspectors have indicated that Contractor could possibly close the door to the room and lock the room out in order to obtain the TCO. HOWEVER, A CAVEAT WILL REMAIN THAT SHOULD A DIFFERENT INSPECTOR PERFORM THE FINALS FOR THE TCO, THIS WORK AROUND SOLUTION MAY BE COUNTERMANDED. CONTRACTOR CAN'T BE RESPONSIBLE FOR THE DECISIONS OF THE COH INSPECTORS. Therefore, should this case arise, Contractor will submit a delay to the substantial completion at such time the appropriate inspections are held, and OWNER acknowledges that this is acceptable. Regardless of the effort to obtain a TCO, Contractor will not have the chair delivered until 10 weeks after approval of the CHANGE ORDER by OWNER. A revision to the substantial completion date will need to be executed at the time of approval of the CHANGE ORDER for the dental chair. Should the CHANGE ORDER not be executed by November 28th, 2016, then additional general conditions will need to be added to the project to cover Contractor overhead.

Major Milestone's Accomplishments:

1. Exterior Skin Complete
2. Elevator Construction Complete
3. Atrium Finishes Complete
4. Gym Flooring installed
5. Commissioning commenced
6. Generator PFC complete

Awarded Delays to Contractor's Contract:

7. Rain Days (15 each) with new substantial completion date of 12/14/16.
8. THIRD PARTY MEP PROVIDER delay with new substantial completion date of 1/7/16.

Attachments:

Project Updated Schedule 11/4/16
Weather Days Report

November 15, 2016

Figure 56: Monthly Schedule Report Page 2.

updating progress, the contractor has failed to communicate project progress in the most expeditious manner and risk is created through lack of communication. To meet the needs of both types of team members, technically savvy and high-level decision-makers, the critical path schedule is distilled into this summary format with the actual detailed gantt schedule attached. In this way, key information is communicated to both type of users and the summary report performs as a periodic snapshot of project progress. Ideally, once the draft of the report has been prepared by the contractor, it should be reviewed by the owner and contractor team so that consensus can be achieved on approved time extensions. Upon agreement by the daily project team, the report can be finalized and transmitted to executive leadership. Following this procedure allows the broadest range of team members to stay informed of project progress. This also allows all levels of ownership to engage in the building process and possible offer mitigation means when the project completion goal begins to slip. Critical to this process is the understanding that the contractor is not solely responsible for having all the answers. The contractor is responsible to provide options for a path forward when the schedule begins to be compromised. The contractor is also responsible to be proactive in their management of the schedule and allow the team time to respond to schedule issues before they impact the project. It has been the author's experience that the best solutions are derived from the entire team working together, since this brings a greater amount of resource application to resolve potential schedule obstacles.

The frequency of the report is generally recommended to occur on a monthly basis since it coincides with financial reporting requirements. However, factors such as project level of difficulty, resource shortages, accelerated completion requirements, should be weighed to determine the reporting period that is best for the project. In the event that these situations occur or others that decrease the probability of completing the project on time, the author recommends that a more frequent reporting period be agreed upon by the team. This accelerated reporting period should remain intact until the project probabilities of timely project completion are stabilized. While this operational move may require more incremental resources to provide, it does focus the team on the time risk elements of the schedule and increases the chances for implementing successful mitigation strategies.

The intangible benefit of implementing this report is the creation of trust between the contractor and the remainder of the team. This is achieved through systematic documentation of the proactive mitigation measures employed by the contractor as the project progresses. Trust can be defined as the belief in the reliability of someone. By consistently reporting on the proactive mitigation measures employed, the contractor demonstrates their focus on the best interests of the project. This in turn, lays the groundwork for trust to form between the owner and contractor. There is one guarantee on a construction project; it will not go exactly as planned. Therefore, it is important to establish trust early in the project as this can make it easier to work through the inevitable elements of change.

5.4.   Part IV: Perfecting Contractual Time Extension Requests

Many contracts between the owner and contractor will have strict provisions on notifying the owner of any extension requests. It's been the authors experience that failure to meet these requirements will nullify any time extension request. And yes, the request can still be void even if the request is justifiable. This condition is a matter of contractual law and is enforceable by the owner. As a result, "perfecting" is a key component when requesting a contractual time extension.

Before embarking on the construction phase of a project, the professionals must know the timing deadline for submission of an extension, and the requirements of content to be submitted. Typically, the time period for submission of the time extension request to the owner is included in the contract. This duration period is usually based upon the start date of the delay until the time of actual request. For commercial construction, the content submission requirements are seldom as regimented or defined as the aforementioned time period of notice. If they are included, then it is imperative that documentation is consistent with contractual requirements or they too can be nullified. If the documentation requirements are not detailed, then standard practice requires that the following be submitted:

Narrative of the delay including the cause, the impacted work, and the impact to the critical path of the project (see Figure 57).
The critical path of the schedule prior to the impact, and
The critical path of the schedule after the impact which includes either the original activities with the delayed durations and/or the additional activities included in the delay.

Exhibit A - Issue Log Narrative

Issue #1: City delay start of Flushing Activities for Water Supply System

Nature of Issue: Time delay outside of control of Contractor

Start of Issue: April 1, 2010
End of Issue: May 7, 2010

Total # of Days: 37

Specifics: The City was performing their Compliance Testing for the EPA and they would not allow the dumping of any chemicals during the testing period. Because of this condition, the flushing activity could not be performed per the baseline schedule. This issue caused the chemical flushing to be delayed, but this also impacted the backfill of the flushing location and the installation of the roadway in this area as well. The owner was informed of this delay in the monthly schedule report for April and May of 2010.

Schedule Reference:
Refer to Activity #'s (Exhibit S): A9000

Figure 57:   Delay Claim Detail Report.

It is always the contractor's responsibility to demonstrate the delay to the schedule, even if the cause of delay is generated by the owner's actions. The narrative is important because it describes the contextual parameters of the delay event. The critical path prior to the delay establishes the mutually agreed upon schedule or baseline. The critical path with the delays then illustrates the impact of the delay in the same format as the mutually agreed upon schedule. Anything less than these three components is considered an incomplete submission. It is critically important to ensure that the delay does in fact impact the critical path. Additionally, concurrent delays will not be accepted by the owner; meaning that two delays will not be granted if they both occur during the same period. Therefore, the contractor should select the longest delay to the critical path when two delays occur simultaneously.

Additional components of a time extension may be required by contract; therefore, it is critical that each contract is examined closely. In some instances, the author has experienced that a recovery plan is a requirement prior to any time extension submissions. In these cases, the owner mandates that the contractor submit a revised schedule indicating how the lost time can be recovered so that the original completion date is maintained.

At this point, focus on the cause of the delay by the owner or contractor becomes of paramount importance in determining fiscal responsibility for recovery actions. Should the owner be the cause for delay, then the contractor is typically within their rights to submit an acceleration schedule with cost impacts to restore the project schedule to its original deadline date. The owner must decide if the acceleration cost is affordable within their budget or if a time extension is acceptable. In the event, the cause for schedule slippage is the result of the contractor, then any costs incurred due to the recovery schedule are borne by the contractor. Most construction contracts are specific with regard to weather delay parameters for a time extension. Weather impacts are generally less ambiguous in determining time extensions to a contract.

In keeping with the concept of risk management, it is then logical to apply a more robust effort in documenting delays which are more ambiguous; regardless of who is the responsible party. During 38 years of experience, the most effective paradigm is providing early notification of delays and document the details of the situation thoroughly. In instances where the contractor fails to provide timely notice, the owner will stand behind the fact that they were not given the opportunity to mitigate the circumstances. The theme of team ownership of the schedule appears again, consistent with the owner's right to approve the baseline schedule. Written documentation is typically required by the contract, but in the absence of this mandate it is always a communication best practice. The author has witnessed conditions where the contractor will wait to submit the notice after all of the facts are obtained. In every instance, this practice puts the contractor at risk. It is common that a delay and the potential mitigation may develop over weeks in lieu of days. Documentation of the delay may take several written forms during the resolution period with the concluding statement consisting of a culmination of all the intermittent notifications.

Another motivation to document issues as they occur is in meeting the test of thoroughness by the information provided. Owner's typically have third-party, or a nonproject team administrator review the time extension request. Therefore, documentation must be illustrative, concise, and effectively communicated so that it's easily understood. During the period of recording a delay or mitigation effort, the goal should be on eliminating questions. Question reduction is typically accomplished by providing more information, not less.

5.5. Part V: Internal Reporting of Schedule Performance

Schedule risk management encompasses both external and internal reporting on a consistent basis. There are many complex measures for internal monitoring of schedule performance, however, the intent of this writing is to present simple and quick measures to ensure timely completion. With those parameters, the author has developed a tracking chart for use on commercial construction projects (Figure 58). This simplified method utilizes common performance indicators:

- Straight line or elapsed time projected progress (diamond line in Figure 58),
- Projected cost percent complete (square line in Figure 58),
- Schedule percent complete per automated software (triangle line in Figure 58), and
- Project manager's percent complete estimate (solid line in Figure 58).

Once the reader understands the meaning of each graph line, then the interpretation of the schedule progress becomes visually discernable. The keys to understanding each data plot line are as follows.

- Elapsed time projected progress (blue diamond line in Figure 58) represents the graphical duration of the project if all work were completed in equal segments by each month. It is important to note that this graph line will not change unless the overall duration of the project is extended contractually.
- Projected cost percent complete (yellow square line in Figure 58) tracks the cost percent complete by each month. The data points to the left of the Update Date Line reflect actual costs and those to the right are projected remaining costs. Most projects follow an "S" curve shape.
- Schedule percent complete per automated software (green triangle line in Figure 58) illustrates the scheduling software percentage completion based upon project updates. These data points are only input as actual values to the left of the Update Date Line.
- Project manager's percent complete estimate (red solid line in Figure 58) demonstrates the construction project manager's subjective judgement of the progress percent complete. These data points are only input as actual values to the left of the Update Date Line.

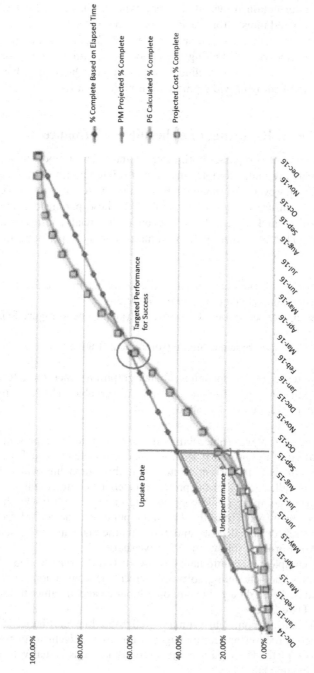

Figure 58:   Underperforming Project Graph.

The evaluation method employs a comparison between the four data points itemized above. One of the most important performance indicators is the intersection point between the elapsed time plot line and the other three plot lines of actual progress of both time and money. In general terms, a successful project will have intersecting lines at approximately 60% complete. In other words, the updated scheduling software percentage completion and the project manager's judgement of percent completion should intersect the lapsed time line at the 60% completion on the vertical axis. It is generally understood that cost completion percentage will lag behind the schedule and project manager's schedule completion data points due to the process of cost incursion after installation. If the two schedule indicator plot lines (schedule and PM projected completion) intersect the lapsed time line prior to the 60% mark, then the project would be considered ahead of schedule. If the plot lines occur after the 60% mark, then the project should be considered behind schedule and mitigation measures should be employed. To illustrate this principle, Figure 58 indicates that the schedule is underperforming based upon the widening gap (cross-hatch area). Immediately following the update date, notice that the projected cost completion line takes on a steeper curve than prior to that point. This would be an indicator that the pace of schedule would need to be increased in order to make the 60% target. The steeper section of the curve will require either more working hours, more personnel, and sooner delivery of materials in order to pick up the

> 60% is the Key Performance Indicator

work pace. In essence, this project is off to a slow start and should instigate mitigation measures to complete on time.

Figure 59 illustrates the opposite condition where the project is tracking ahead of schedule. In this example, the computer scheduling software (P6), is the top line of the graph on a consistent basis and represents the project updates on a monthly basis. The gray cross-hatch area is the overperformance margin when compared to the straight-line calculation of the project's overall duration. It is rare to see a project consistently tracking ahead of schedule and caution would predicate that a closer analysis of the schedule be performed. Specifically, the content of the schedule should be reviewed to ensure that adequate detail has been included for the remaining activities, beyond the data date. Additionally, previous updates of the schedule should be reviewed for realistic reporting of progress. These two tests should validate the validity of schedule tracking represented in the graph.

Figures 60, 61, and 62 are from the same project but display different information. The intent of including all three figures is to show the correlation of time delays to varying project resources. Figure 60 demonstrates the timeline of actual construction events as well as the progress compared to the straight-line graph. Of particular note, in July 2013, the project experienced a drop-in schedule productivity. This is directly due to a major delay with unforeseen underground utilities and the addition of tenant buildout of a previously programmed shell space. The following month, a recovery schedule was implemented in an effort to mitigate both unforeseen issues. Further examination of the chart shows

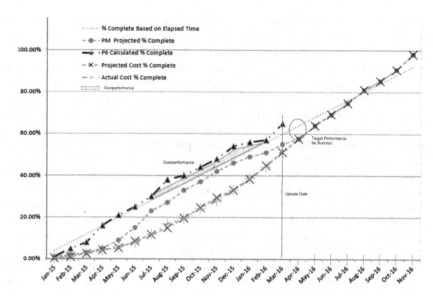

Figure 59:   Over Performing Project Graph.

that the (P6) schedule software updates never intersected the straight-line graph. This means that the project was in a perpetual state of being behind schedule. The data line on July of 2014 reflects the actual substantial completion date and final completion occurring the following month.

Review of Figure 61 shows the relationship of the original vs current cash flow projections. Notice the drop-in project cost in July 2013 corresponds to the schedule delays described above. The steep slope line of the months from August 2013 through December 2013 are commensurate with a stacked resource schedule. Stacking resources is commonly called crashing the schedule due to the increased workers over a short duration. The positioning of the current cash flow line to the right of the original projection also mirrors the delay in the schedule.

The last Figure 62 illustrates actual vs projected labor units per the project schedule. The dark bars represent the actual hours completed per the schedule updates and the lighter bars are the projected labor hours. The month of May 2013 shows significant underperformance in work hours which correlates to the significant shortage in actual cost in Figure 62.

Figures 61 and 62 can be characterized as micrographs of cost and time components that are included in Figure 60. For time management purposes, the summary graph illustrated in Figure 60 can provide a quick and visual management trend of the project schedule. As discussed in Chapter 1, change is inevitable, and the graph provides a visible method for recognition. Monitoring the data points in Figure 60 on a regular periodic basis facilitates earlier interventions. The probability of successful risk management is increased when mitigation measures are implemented earlier in the project.

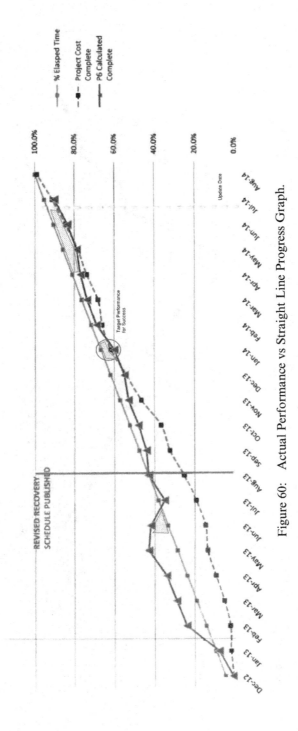

Figure 60:   Actual Performance vs Straight Line Progress Graph.

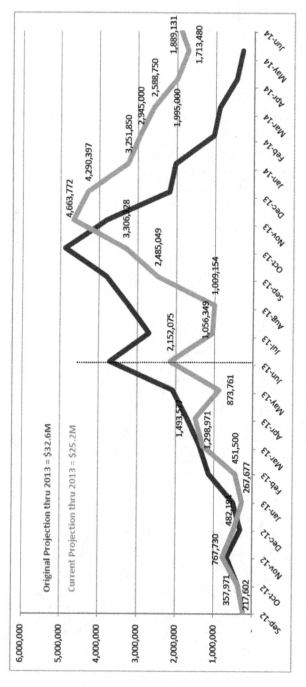

Figure 61: Cash Flow Variance Graph.

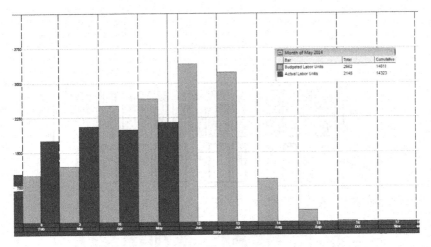

Figure 62: Manpower – Actual vs Projected Graph.

5.6. Part VI: Close-Out Executive Reporting

Earlier chapters discuss the importance of presenting the schedule in an intuitive format. An easy to read format serves two purposes. For the schedule-adept team members, this simplified format enables a drill down effect into more granular details without sorting through activities which are not critical to completing the project. For those team members who are not trained to easily read a gantt schedule, a simpler format removes the learning curve. Lastly, a summary report allows the critical milestones to be emphasized with quick identification so that team discussions are facilitated regarding successful execution. This approach is particularly important at the end of the project when the highest numbers of people are working to complete and activate the project. It's at this point in the project when nonconstruction-related team members typically join the project who are involved in building security, furniture, and information technology system installation. Additionally, commissioning personnel have also been added to the team for the purpose of verifying the functionality of occupancy systems in the building. With the infusion of more people to the project team and the typical shortened time period to perform activation, speed of execution becomes the paradigm. This condition accentuates the need for simplified and effective communication. It is with these core understandings that the activation report in Figures 63 and 64 were designed by the author.

> **Micro Reporting Can Increase Focus at Key Moments of Risk in a Project**

The content focus for the reports in Figures 63 and 64 includes any activity that can impact the occupancy date by the owner. At this phase in the project, the contractor should be working to establish and complete the punchlist for the project. Hence, the report includes tracking of punchlist completion in smaller areas of the building. By breaking down the overall project into smaller turnover

ACTIVATION TARGET DATES

Level	% of Floor Punched	Punchlist Count	Remaining Punchlist Count	Punchlist % Complete	Target Date of 100% Complete Punchlist & Signoff	Furniture Activation Start	Furniture Acceptance	TDH Semi Final Inspections	IT Desktop Installation	Earliest Staff Occupancy/Move	Issue Status
B2	100	223	-	100%	6/18/10			5/3/10	6/24/10	5/23/10	G
B1	100	45	-	100%	6/18/10		6/24/10	5/3/10	6/24/10	7/8/10	G
L1	100	331	-	100%	6/18/10	6/22/10	6/24/10	5/3/10	6/24/10	7/17/10	G
L2	100	258	-	100%	6/18/10	6/22/10	6/24/10	5/3/10	6/24/10	7/13/10	G
L3	100	125	-	100%	6/18/10	NA	NA	4/12/10	NA	5/23/10	G
L3.5 - 12.5	100	1,131	-	100%	6/18/10	NA	NA	4/12/10	NA	5/23/10	G
L13 (16)	100	1,134	-	100%	6/18/10	6/17/10	7/6/10	5/3/10	7/7/10	7/6/10	G
L14 (17)	100	1,327	-	100%	6/18/10	6/17/10	6/22/10	5/3/10	6/22/10	7/9/10	G
L15 (18)	100	251	-	100%	6/18/10	6/17/10	6/22/10	4/19/10	6/22/10	7/6/10	G
L16 (19)	100	735	-	100%	6/18/10	6/17/10	6/22/10	4/26/10	6/22/10	7/9/10	G
L17 (20)	100	648	-	100%	6/18/10	6/4/10	6/17/10	4/19/10	6/17/10	7/13/10	G
L18 (21)	100	753	-	100%	6/18/10	5/13/10	6/4/10	4/19/10	6/4/10	7/13/10	G
L19 (22)	100	666	-	100%	6/18/10	5/11/10	5/20/10	4/26/10	5/20/10	7/13/10	G
L20 (23)	100	656	-	100%	6/18/10	5/3/10	5/14/10	4/26/10	5/14/10	7/7/10	G
Bistro	0	0	0		8/5/10	?	?	8/5/10	?	8/5/10	G
L21 - 23 (24-26)	100	337	0 #	100%	6/18/10	NA	NA	4/26/10	NA	5/23/10	G
		8,620	-	100%							
				100%							

Figure 63:   Construction Activation Executive Summary Report.

	Target Date 100% Complete	% Complete	Issue Status
OWNER ACTIVATION BY OWNER STAFF			
FMS Monitor BAS	7/1/10	100%	G
FMS Monitor Ptube	7/1/10	100%	G
Security Access Control	7/18/10	98%	G
FMS Monitor Med Gas	7/1/10	100%	G
IT Network Operational	7/18/10	98%	G
Fire Alarm Monitoring	5/28/10	100%	G
Keying to End Users	7/18/10	95%	G
Housekeeping Operations	7/6/10	100%	G
Dock Operational	7/19/10	80%	Y
Parking Garage Operational	7/19/10	80%	G
Valet Operational	7/13/10	90%	G
Insurance Activation	5/21/10	100%	G
Food Service Operational	7/13/10	60%	Y
Temp Security Monitoring	5/21/10	100%	G
Final Security Patrol	7/10/10	90%	G
Supply Chain Operational	7/1/10	100%	G
Conference Room Services	7/19/10	95%	G

Figure 64:   User Services Activation Executive Summary.

units, the construction team can more easily manage the process and allow the owner to take partial occupancy of the building. This phased approach for occupancy can facilitate a shorter overall duration for the activation period. Activity milestones for furniture commencement and completion as well as technology "go live" dates provide a dashboard for multiple department managers. Estimated move dates for the buildings inhabitants is also include in the next to the last column. The last column allows a subjective assessment of the overall process by row. This assessment is essential for effective management of the complete process in bringing the project to completion. Utilization of a "red/yellow/green" designation of the box facilitates a familiar visual notification of projected status. Hence, should the manager who originates the report determine that more action is necessary, but not critical, a yellow designation would be applied. If the situation needed major problem-solving work to generate solutions, then a red designation would be more appropriate. Green indicates that all activities for this area of turnover are proceeding as planned. While not illustrated in the figures of this chapter, the author has successfully employed this technique with the commissioning process on large projects. The items included for tracking can be easily modified through the use of an excel spreadsheet to fit summary milestones for the commissioning process.

Figure 64 focuses on the activities by owner departmental groups which must be completed prior to building ownership. This report is summarized at a high level to serve as an executive report. The items included for tracking can be easily modified through the use of an excel spreadsheet to fit the owner's needs. By focusing on these support operations to the building, the project manager can ensure successful turnover. While this may appear to be content particularly intended for the owner, it's important for the contractor to understand that if the owner is not ready to take occupancy this will cause additional general conditions expenses. Depending upon the terms of the contract, this condition could result in project savings erosion.

From the owner's perspective, construction is a means to an end condition. Therefore, a good building partner is focused on the ultimate needs of the owner. Alteration of the contractor's perspective to align with this concept creates the best condition for a strong partnership with the owner. It has been the author's experience that the owner, architect, and contractor team always remember how the project is finished, not necessarily how it was started. This reality implies the emphasis on completing the project with a reporting system that enables problem-solving and timely completion by all parties.

Chapter 6

Contingency Planning

The predominant theme of this book has been that change is inevitable. Successful projects typically have a process and resources in place to deal with change. The deliverable of this process is commonly referred to as the contingency plan. In simplistic terms, a contingency is a plan focused on an event that is in the future and cannot be predicted with certainty. It would be exhaustive if every risk had a developed contingency plan, therefore the author's practice has been to generate contingency plans for only the highest risks of a project. The collaborative approach between project team members will typically reveal those areas of highest risk. Once those risk items are determined, then development of the contingency plan is recommended. Of equal importance during this process is the determination of the tasks which need to be tracked to monitor progress of the original plan. The team must also determine the optimum date in which to implement the contingency plan.

6.1. Part I: Understanding the Difference Between Contingency and Mitigation

Another common term in construction practice which is similar to contingency is mitigation. It is important to distinguish the difference between mitigation and contingency. Mitigation is the action taken to amend a situation to its original completion date. Based upon this understanding, mitigation measures

Mitigation Measures Minimize Impact

can be components to the contingency plan. This definition implies that mitigation happens after schedule slippage. Industry contractors typically refer to actions to reduce time erosion during normal work processes as mitigation, regardless of the pre- or post-active nature of the impact. The timing of the effort is a small point of distinction. The primary importance of mitigation is that it decreases the duration of an activity by employing proactive steps. Examples of common mitigation measures are as follows:

- Avoid handling materials more than once.
- Adjust crew sizes so that they can work effectively in the work area.
- Determine the origination point of trades working in the same space either as singular or oppositional starting locations.
- Working double shifts with work crews.

- Use of productivity analysis on specific repetitive types of work (i.e., precast erection, tilt panels, etc.).
- Efficient arrival processes for workers and materials to the jobsite.
- Minimizing safety downtime due to accidents.
- Minimizing rework or corrections of initially installed work.
- Utilization of the use of near and long-term planning during construction execution is a disciplinary step to maintain focus on productivity and installation.
- Restrictions for change in scope of work past a certain point in the project to allow completion of existing scope.

Further elaboration on some of the mitigation measures is warranted in order to achieve complete understanding. Specifically, handling materials more than once is a subtle means for time erosion and can have a large impact on the schedule. Ideally, materials should be located on site adjacent to the installation area upon arrival from the manufacturer. This creates an optimum condition where the distance the worker has to retrieve and install the material is minimized. For instance, sheetrock material should be located on each floor of a multistory project in lieu of in one central location on one floor. This approach minimizes travel along the vertical passageway. In the case where a floor size is large, then consideration to spread the sheetrock material bundles out closer to the installation area is recommended. This will reduce actual material handling time and minimize fatigue by the workforce. For example, in a large floor plate of approximately 65,000 sf, overall material handling time can be reduced by placing several stacks of gypsum board closer to the installation space. In the case where a smaller floor sf exists (i.e., 10,000 sf), this may not be physically possible. When assessing the project site's capability for this type of material management, it is important to consider mobility requirements of machinery and personnel before finalizing the material management plan.

Crew sizing by various subcontractor trades in the work space is important because productivity losses generally occur when too many workers are in the same space. The flow of work crews in a work space can also impact time loss. Work flow can occur in a singular direction or as "ships crossing in the night" from opposite origination points (Figure 65). By starting from opposite sides of a work area, this allows more work to commence simultaneously with the only potential slow-down period occurring when the crews meet in the middle of the space. This methodology is also effective when starting dissimilar work crews from different origination points. The singular direction method will not gain as much time on the schedule since each trade crew must wait to start until their predecessor is sufficiently progressed in front of them.

Working double shifts is generally considered a nonpreferred method to mitigate the schedule. There have been numerous productivity studies conducted which show that time gain from implementation of this measure is not a one-to-one return. Worker fatigue by working experienced during nonregular hours will silently slow productivity. This method should be considered carefully and used only when partial productivity is considered acceptable.

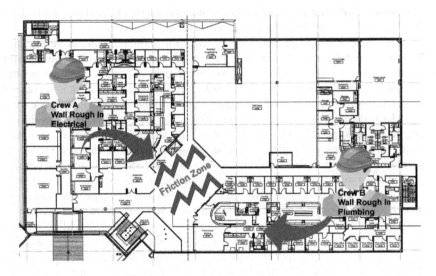

Figure 65: Work Flow Mitigation Method.

Worker access planning is just as critical as material staging to the work area. Time lost due to installers waiting for a lift to upper floor work areas can be significant, particularly during shift starts and completions. It is important to assess vertical floor access either through stair or mechanical lifts. This strategy must be examined with an understanding of the volume of workers utilizing the access ways. Parking availability to the site also has the potential to reduce precious time that can be utilized during construction processes. Urban sites pose a risk of time loss due to worker transportation to the site prior to construction activity commencement each day.

Safety is an integral aspect of construction processes and can have a significant impact on the construction schedule. The author has seen jobsites where an increase in work hours and multiple shifts produced an increase in accidents. In the condition where workers are not adhering to proper safety procedures, safety standdown's can increase which results in a cease in work production. As described above, safety incidents can create compounding time loss to the schedule. The intangible impact of multiple safety incidents can create a tentative workforce which can erode productivity. The effort to make up lost time by pressing the workforce requires a delicate balance. Typically, this mitigation measure is best utilized for short burst of time vs long-term work policy.

Obviously, reworking an installation only acts as an erosive element to the schedule since the construction is performed twice. There is typically a period of time that would be added to the effort based upon removal of the incorrectly installed construction components. Constructability planning can greatly aid in the reduction of rework based upon the preparedness of the workforce through the process. Advance planning and crew understanding of the scope of work to be installed is a critical key to success with constructability planning.

Near- and long-term planning are essential for maintaining focus during construction operations and adjusting timing of execution based upon natural changes during the building process. Near-term planning is implemented through the use of three-week look ahead schedules where the work is scheduled for each trade on a daily basis. It is important to note that the three-week look ahead schedule only includes those activities which are projected to occur during the three-week period. The master or parent schedule includes all activities required to complete the project and serves as the baseline reference for timely execution. The three-week look ahead schedule is generally utilized in weekly project meetings with the foremen from each trade. The near-term schedule is considered the most granular or microscopic version of schedules. The look ahead schedules should be coordinated in tandem with the master schedule or long-term plan. Specifically, the near-term schedule activities should fall within the duration parameters of its parent master schedule. Oftentimes, the contractor will work to shorten the durations of the look ahead schedule for the purpose of creating float relative to the master schedule. Building float as construction proceeds is highly recommended as it provides a hedge against unforeseen delays which will typically impact a schedule. The look ahead progress schedule is then used to update the master parent schedule with actual and projected completion dates. Typically, these updates occur at a minimum on a monthly basis. Each project should be evaluated on the risk or difficulty of execution to determine the best suited periodic update duration. The discipline of managing these schedules through these practices enables early detection of schedule erosion, which increases the probability for mitigation. In the case of creating float, these practices create float as a contingency for later delays. In order to obtain maximum effectiveness of mitigation measures, they must be applied as early as possible in the project. This mindset allows the opportunity for change adjustments as the project develops.

One of the last methods and probably the hardest to implement is restricting scope changes beyond a certain point in the project delivery period. This typically would occur during the last quarter of a project's overall duration. The hardest part is saying "NO" to your client. A contractor's number one goal is to complete the project on time, within budget, and with a happy client. The impact of

> **Responsiveness to Scope Change is Key to Success**

using the "NO" word may lead to an unhappy client. However, the author has seen projects where too many changes occur during the last stretch of the project and there is either not enough space for all the workforce or simply not sufficient time to complete the changes. Clients depend upon contractors for their honest and timely assessment of the construction process. While the short-term result may not be the optimum, an owner will appreciate the ability to open their facility on time.

These methods are, but a few of those, that can be employed in an effort to conserve work days for future surprise impacts. Understanding the importance of disruptions to the work flow is a major step in preventing time erosion to the schedule. From an esoteric perspective, construction is about integrating smaller pieces into the whole. The construction manager must always be attuned to how each operation impacts the overall production schedule. Adaptability is the most critical characteristic of schedule management. Mitigation and contingency planning are the means of implementing flexibility during the construction process.

6.2. Part II: Reporting a Contingency Plan

A contingency plan in simplistic terms is a "what if" plan in the event that the schedule slips beyond the baseline critical path. These plans are often developed for the top percentile of most likely delay events during the project process. In other words, the chances of a delay for this activity or process is high. Since contingency planning is similar to forecasting

> Contingency Planning Content = Highest Risk
> Operations

with a crystal ball, and can be time consuming, the number of contingency plans will typically be small in nature. Another characteristic of a contingency plan is that it is developed in advance of the calculated delay. Elements to be included in the contingency plan should be as follows:

- Description of the risk factor which could cause the delay,
- Mitigation measures to restore the critical path to the baseline,
- Appropriate emergency actions associated with the impacted event,
- Timing for implementation of mitigation measures,
- An executive summary for leadership personnel, and lastly,
- Microdetails for the crews performing the work.

Actual contingency plans are illustrated in Figures 66, 67, and 68 which demonstrate the components described above. Figure 66 specifically addresses the emergency actions in the event of severe weather, and utility outages that may occur as a result of the demolition/implosion operations. In this particular event, the elements of risk are significantly higher based upon the destructive operation. This summary overview facilitates efficient communication of risk procedures to a broad audience.

Figures 67 and 68 illustrate a summary and detailed description of the contingency operations for the same event, respectively. The implications and impacts are critically important in communicating the potential risk of the operation. Careful review reveals that the use of a generator is employed during the shutdown to feed active campus operations.Should the ongoing campus operation be determined to be mission critical, then the use of a secondary generator may be considered as a means to safeguard against potential contingency failure. The abbreviated nature of this executive summary report allows key decision-makers for all entities involved to assess if the contingency plan is sufficient to meet the ongoing needs of the operating facility.

Implosion Contingency Plan

a. Debris in Holcombe or Fannin or Roadway Damage

The City of Houston traffic control plan will remain in place until the debris is cleared and the roadway is repaired and re-opened.

b. Severe Weather

Lightning directly overhead, or a heavy rain or snow obstructing the view of the structure will delay the implosion until the disturbance passes by. In the event that high winds bring down building protection such as geo-tex or plastic the countdown will be stopped and will not restart until the protection is reinstalled.

c. Damage to Adjacent Institution

All surrounding institutions have attic stock glass available for use in the event that a window is broken. The demolition team will have lumber, sheathing and plastic on-site to temporarily infill broken windows or roof areas until a permanent repair can be made. The demolition team will have a roofing and glazing contractor on standby during the implosion.

d. Power Outage

Centerpoint Energy will have a representative in the command post during the implosion that will have contact with his service crews outside of the exclusionary zone. The repair crews will be staged at the corner of Braeswood and Fannin and will be allowed access within the exclusionary zone immediately following the implosion to begin inspecting surrounding power lines and the Pressler Street substation.

e. Water or Gas Outage

The City of Houston water department and gas department have been notified of the implosion and will have crews on standby to make repairs to damaged water or gas lines, if necessary.

Figure 66:   Emergency Action Contingency Plan.

Figure 68 represents a portion of a detailed contingency plan where the specific activities are described as well as personnel and equipment. This level of reporting is important to the personnel who are responsible for implementing the plan. Therefore, microscopic detail is required for each activity, timing, personnel, and scope of contingency measures. For the purpose of this manuscript, only the first page of the report is included in the illustration. The thoroughness

---

Electrical Shutdown Summary and Schedule

1. **CPDP Panel - Central Utility Plant (CUP) Electrical Distribution Panel**
   a. Shutdown scheduled for 3/9/18 – 3/11/18 (48 hour)
   b. Shutdown implications
      i. Chiller 1
         1. Disconnected from main power feed.
         2. Will be fed from existing HMH generator
      ii. Chiller 2
         1. Disconnected from main power feed.
         2. Will be fed from secondary generator package
   c. Noticeable Impacts
      i. None.
      ii. Both Chillers running normal on generator power.
      iii. With both Chillers being back fed by separate generators multiple redundancies in place if possible issues should arise.
      iv. With the forecasted weather for this shutdown one chiller will be able to supply all required chill water
   d. See attached approved shutdown plan for reference
2. **MSB Panel – Normal Power to Hospital Electrical Distribution Panel**
   a. Shutdown Scheduled for 3/17/18 – 3/18/18 (7 hour)
   b. Shutdown implications
      i. Normal power (white receptacle & Lights) interruption
   c. Noticeable Impacts
      i. Minimal.
      ii. Similar to normal power outage.
      iii. Maximum 11-second delay for generator start-up.
      iv. HMH Generator to feed normal power during shutdown.
   d. See attached plan of action, final signoffs expected on 3/9/18.

Figure 67:   Executive Summary Contingency Plan.

of the plan is especially important, since most contingency operations typically occur during nonregular work hours. Advance communication becomes paramount during this off-cycle work period.

In commercial construction, there is a distinction between contingency planning, recovery planning, and acceleration. Each of these three plans share the same elements of a contingency plan, however, the timing relative to the delay and the cost for mitigation measures will vary (see Table 4). Typically, acceleration plans are associated with financial responsibility being born by the owner, where the owner's actions have created the delay. In this case, the contractor will prepare a plan which includes the cost to provide compression of the schedule which will result in recovering the overall duration to the original baseline. Based upon the timing of this effort, this type of plan is typically developed after the delay is incurred. Contingency planning is considered as a proactive measure to avoid delays, where they can be employed prior to the actual slippage in the schedule. If the contingency plan creates additional cost, then that is born by the contractor in an effort to deliver the schedule per the contractual terms of agreement. Another common commercial construction term is recovery schedule. Basically, this is a post-delay plan where the contractor has failed to implement

Activity: Shutdown switchboard "CPDP" to test generator can handle Load for 12+ hours before CPE shuts down CPDP XFMR. Once generator has been verified, CPE will shut off XFMR for us to start our work as follows - install new breaker, conduit & conductors to feed new panel "PHEQDP" & "WIREWAY NEMA 3R". Tie specified equipment into Temp/Existing generator that will run for the duration of the shutdown. Everything will be verified with Representative at the Hospital before we start the shutdown.

Date of Shutdown:

Start Date: 03-09-18

Duration: 48 Hours - 3/09/18 - 7PM to 3/11/18 - 7PM

Crew information:

Crew Size	Duration	Task
10	3-9-18 4 Hours – 7PM–11PM	Disconnect and Reconnect existing equipment and tie in generator leads
5	3-10-18 12 Hours – 7PM–7AM	First shift in CDPD
5	3-11-18 12 Hours – 7AM–7PM	Second shift in CDPD
10	3-11-18 4 Hours – 4PM–7PM	Disconnect and Reconnect existing equipment and tie back into existing system

Electrical Equipment Shutdown:
- Switchboard "CPDP"

Mechanical Equipment Shutdown:
(SEE ATTACHED ONE LINE)

Temporary/Existing generator prep work to be complete before shutdown
1) Lay out Generator cables.
2) Set up Temporary switchboard
3) Deliver temp generator to site and park behind ambulance canopy.

Figure 68: Detailed Contingency Plan.

Table 4: Schedule Delay Planning Matrix.

Form of Schedule Plan	Timing Relative to Delay	Cost Responsibility
Contingency	Prior	Contractor
Acceleration	Post	Owner
Recovery	Post	Contractor

a contingency plan and the cost of schedule restoration is born by the contractor.

In summary, a contingency plan can be developed for any situation. The topic of the contingency plan should be carefully considered to include the highest risk elements with the greatest probability of nonplanned performance.

Knowledge gathering, and creativity are instrumental to the development of an effective plan. In order to maximize the float or "time contingency" in a schedule, it is best to create these plans earlier in the project, rather than later. This proactive approach allows the project team to make more adjustments which increase the probability of success. The intangible benefit of early monitoring of the schedule and contingency plan development is creation of trust between the owner and contractor. The strength of the relationship between the owner and delivery partners cannot be undervalued. All projects will have delays at some point in the process and the ability to overcome is based upon planning and mutual trust that all parties are in control of the situation. The tangible benefit of contingency planning is either to shorten or maintain the original baseline schedule. Acceleration and recovery plans are directed at mitigating the delay and restore the critical path to the original baseline schedule. Regardless of which form of contingency plan is utilized, they are sound tools to increase the percentages of completing a project on time.

Chapter 7

Implementing Mitigation Measures Through Technological Application of the Computer Scheduling Software

7.1. Part I: Technical Parameters of Activities which Reduce Risk

Until now, the manuscript has dealt solely with the stage of preparing a schedule prior to utilization of electronic software. This chapter will illustrate key implementation techniques when inputting the content into a software program. Each of these concepts have been developed for the purpose of minimizing risk in the project schedule and are as follows:

- Activities in the schedule should have durations of nine days or less.
- The schedule should ideally only have one critical path.
- Total float should be designated at zero days.
- Free float between activities should be within the range of 0−50 days.
- Weather days for interior scope of work activities can be consolidated into a single activity titled "schedule contingency."
- Predecessors and successors logic should be applied to all activities except for the predecessor to the first activity and the successor to the last activity, and
- Work breakdown structure (WBS) codes should be applied to each activity and more complex projects should consider utilizing activity codes.

The purpose of maximizing the duration of single activities to nine days is to provide a granular approach to the schedule. Typically, the more detailed definition in activities yields a better plan for execution. The purpose of a singular critical path provides simplicity of focus on the most time sensitive activities in the schedule. The importance of inclusion of activation activities is that they are typically controlled by the most outside resources. The contractor does not always have a direct contractual relationship with these vendors, so they fall within a higher risk category. With regard to the critical path, most computer software programs have the capability of modify the total float duration. However, a zero duration for total float is typical industry practice. The purpose of maximizing the free float between activities is similar to the reasoning for limiting activity durations to nine days or less. Larger gaps in time between activities can reflect less planning of the sequencing of events. Consolidation of the interior activity weather days into a single contingency line item allows the schedule reviewers to focus on the use of float at each update period. This practice is easy to justify especially since interior activities do not incur weather impact. The purpose of predecessor and successor logic to all activities is key for creating a tight schedule. "Dangling" activities, aka nonlinked, lack logic and can create opportunities for time erosion. And lastly, the power of scheduling software is the ability to mine data sequences for problem-solving when activities do not progress as planned. As this manuscript has illustrated, the probability of a baseline schedule progressing as planned is minimal, so the necessity to evaluate adjustments in sequencing and durations is critical to maintaining the original project duration. Since WBS and activity codes for each activity enable prompt and accurate data identification, their use is of paramount importance.

7.2.   Part II: Mitigating Mistakes during the Schedule Input Process in Computer Software

All experienced practitioners know that the work day is continually interrupted by the activities on the job site. As a result, periods of focus time on schedule-building are seldom available. Therefore, it is important to have a system which mitigates these breaks of mental concentration. The author has developed a system during the last quarter of a century which maximizes the ability to focus on schedule input in spite of multiple interruptions. These proven techniques enable your mind to work in one direction, so when you are distracted from the schedule development process, it is easier to resume creation. There are four basic phases during this sequence of operations:

(1) initial schedule development;
(2) evaluate and adjust;
(3) format, save, and publish.; and
(4) updating and responding to change.

The phases and the steps described below are the key to consistency and accuracy in the creation of a quality schedule. The most important aspect to the process is to complete the steps in order. This facilitates performing a repetitive task and enables easy identification of where you stopped at the time of the distraction. From a focus standpoint, it is easier for the brain to focus on one type of activity at a time. For example, it is easier to list all the activities before moving to the next function of assigning durations versus listing a singular activity, it's duration, then repeating the process for the next activity. It has been the author's experience that this type of repetition also makes it easier to re-engage in the schedule-building process. The recommended steps for schedule creation and maintenance are listed in the form of a checklist below.

- Phase 1: Initial Schedule Development
  - Step 1 – Set-up the parameters in the scheduling software for weather days per month, work calendar days, project start date, WBS structure, and critical path duration in days. (Note: these steps are specific to P6

software. Further research may be required for the exact parameters with different software each will have the same generic categories of information).

— Step 2 — List all activity names. In order to avoid assigning WBS codes to each activity, the author recommends that activities are added to each WBS section of the schedule. This method duals as an intuitive process since listing activities for similar operations, i.e., concrete placement vs exterior skin construction, is an easier focal point.

— Step 3 — Assign durations to all activities. This step is as simple as it sounds. Now that all activities have been listed, the process moves to contemplation of the time to complete each activity. By focusing on this only aspect of the activity, the brain can get in a rhythm of thinking that will facilitate speed in the process. This method also allows easier identification of where you left off in the process after an interruption.

— Step 4 — Link predecessor and successors for all activities. Predecessor is the activity which must come before a specific activity and the successor will follow. There are four specific relationships for predecessors and successors, and this book does not delve into depth on these aspects. Other textbooks can elaborate on the intricacies of the character of the relationships that can be defined between activities. For the purpose of this book, it is important to know that they are as follows: FS — finish to start, SF — start to finish, SS — start to start, FF — finish to finish. Step 4 should include assigning both predecessors and successors to all activities in the schedule as well as one of the four relationships. By establishing all of these predecessor/successor relationships at once, the thought process is more simplistic and repetitive which facilitates efficiency.

— Step 5 — Calculate schedule. This is the process where the computer algorithm takes the information provided in steps one through four and generates the critical and noncritical paths of the schedule.

The key to all Phase 1 steps are to complete them before moving to the next step. In simplistic terms, all activities should be listed in the software before moving to the assigning of durations. This method of data input allows the brain to focus on one function at a time and makes it easier to return to your task after a disruption in the process. Once the schedule has been calculated, the process of preparing the schedule is only one-third complete. The biggest mistake typically made at this phase is to rush to publication. The next phase is essential for confirming the integrity and accuracy of a schedule.

- Phase 2: Evaluate and Adjust
  — Step 6 — Use the critical path filter to show only the critical path activities.
  — Step 7 — Review the critical path for inclusion of all appropriate activities
     These activities are identified with 0 float or the lowest amount of float in the schedule. If there appears to be some activities missing from the critical path, that is, elevator, permanent power, etc. then add or link the

missing activities to activities shown on the critical path. Lastly, rerun the filter to review again, until the proposed critical path accurately reflects project risk. Once this step is complete, remember to return the filter function to show all activities before moving to the next step.

- Step 8 − Confirm that all activities have both predecessors and successors with exception of the first and last activity in the project schedule. Most software applications have reports that will provide this information. If that is not available, then a predecessor and successor column can be added to the gantt chart schedule for easy identification.
- Step 9 − Evaluate the total float for each activity so that durations are not too lengthy. This process is commonly referred to as tightening-up the schedule. Lengthy durations in the total float column can also indicate missing logic relationships. The net effect of this condition can create late notifications of impact to the critical path. The goal is to have a tight schedule with all activities logically linked together so when an activity is delayed on the critical path, the scheduler is notified of the true progress of the job.
- Step 10 − Confirm that the owners WBS section of the schedule includes the notice to proceed milestone, the contractual duration of the entire project, and the substantial completion. Then confirm that the Projected Substantial Completion is included and linked to a milestone in the same owner WBS (see rectangular boxes of activities in Figure 69). This is important as it includes the legal milestone dates from the contract onto the schedule for easy reference. For easiest and best use of this schedule function, it is recommended that this WBS section be shown on the top of the first page of the master schedule. This facilitates a quick review by the users when reviewing the schedule updates. Figure 69 is an example of where this key information is provided on the first page of the schedule.
- Step 11 − Review the schedule for stacking effect. Stacking can happen to activities when the schedule is compressed in an effort to mitigate a delay. In simplistic terms, stacking is where multiple activities occur at the same time. The stacking effect is illustrated within the box located in Figure 70. The impact of this effect can be labor shortages which increase risk of inability to meet the schedule. This condition should be avoided. If aversion is not possible, then labor measures should be evaluated to increase the probability of meeting activity deadlines.

- Phase 3: Format, Save, and Publish
  - Step 12 − Save baseline. At this point, the schedule is ready for publication. The baseline for the original master schedule and any subsequent updates should be saved in the computer application. By saving at these snapshot points in time, an audit trail is created that can be accessed to explain changes in the schedule.
  - Step 13 − Confirm the layout view for the schedule will display the baseline and progressed dates for each activity. The purpose of this action is to

Owner Contract Milestones- ED		387	05-Aug-16 A	18-Sep-19
ED-M10	Notice to Proceed Preconstruction	0	05-Aug-16 A	05-Aug-16 A
ED-M32	Proposed Contract Modifications Sent to HMH	0	15-Aug-16 A	15-Aug-16 A
ED-M64	Legal Negotiations on Contract	0	15-Aug-16 A	24-Aug-17 A
ED-M52	HMH SJ Finance Committee Approval of Budget	0		17-Nov-16 A
ED-M62	HMH Board Approval of Budget	0		30-Nov-16 A
ED-M72	PSA ED Preconstruction Rcvd by Tellepsen	0	30-Nov-16 A	30-Nov-16 A
ED-M42	Target Need Date to Facilitate Make Ready & Early Mobilization	0		06-Mar-17 A
ED-M44	Notice To Proceed - Make Ready Canopy ONLY	0	09-Jun-17 A	
ED-M14	Notice To Proceed - Make Ready and Temp Parking Lot - A	0	28-Jun-17 A	
ED-M18	Mobilization Make Ready - A (BASED UPON PERMIT RCPT)	0	14-Jul-17 A	
ED-M20	Contract Construction Duration	547	14-Jul-17 A	18-Sep-19
ED-M16	Notice to Proceed - ED Renovation - C	0	01-Nov-17 A	
ED-M12	Notice to Proceed ED Expansion - B (1st Floor)	0	01-Nov-17 A	
ED-M82	Notice to Proceed - ED Expansion B (2nd & 3rd Floor)	1	01-Nov-17 A	07-Mar-18
ED-M34	CPE PowerPoles & Ancillary Service Removal Along Baker Rd (By Owner)	0		01-Nov-17 A
ED-M54	Telecom Relocation of Service by Owner Due Date	0		07-Nov-17 A
ED-M22	ED Expansion Substantial Completion - B (Level 1 Only)	0		15-Mar-19
ED-M24	ED Renovation Substantial Completion -C	0		18-Sep-19

Figure 69: Owner WBS Schedule Section for Contractual Terms of Time Progression.

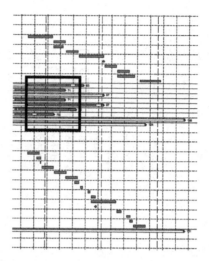

Figure 70:   Schedule Stacking Effect.

	ACTUAL START	REMAINING DURATION	ACTUAL FINISH
Activity Description	✖		✖
- OR -			
Activity Description	✖	✖	

Figure 71:   Project Activity Update Sequence.

allow visible identification of each activities progress relative to the agreed upon baseline. This is an intuitive means of quickly assessing whether a schedule is ahead or behind.

–   Step 14 – Assign activity codes for each activity. This step is optional for projects which are relatively simple and straightforward. Complex schedules should utilize the functionality within scheduling software. This "layered" coding of each activity enables data query within the schedule based upon the activity codes employed. As previously discussed, this is recommended when the contractor wants to customize the schedule to job specific attributes.

7.3.   Part III: Mitigating Errors during the Update Process

- Phase 4: Updating and Responding to Change
  - Step 15 – Update the schedule. Figure 71 illustrates the data needed to update the progress of individual activities within the schedule. Constrained dates should never be used to update activities as this limits the software's ability to retain the original schedule logic.

- Step 16 – Implement the critical path filter to show only the critical path activities. This allows the reviewer to precisely evaluate the impacts to the critical, aka. longest, path of the schedule.
- Step 17 – Confirm that the substantial completion has not changed from the current contractual completion date. If the critical path becomes longer than the contractual date, then the contractor needs to re-evaluate either activity durations or logic to bring the schedule back to the agreed upon baseline completion. If there are legitimate delays that qualify for a contractual extension, then the scheduler should move to Step 18 and input the delay synopsis and time extension request.
- Step 18 – Add delays and weather impacts in the summary owner WBS section of the schedule. Notice to the owner is a required fundamental when attempting to perfect a delay claim. The contractor is responsible to provide the opportunity to mitigate the delay whether through their or other team member means. Hence, the importance of providing proactive information on the potential for a delay. The updated activities will reflect the delay, when incurred, and it is equally important to accentuate the reasoning for the delay. As a result, the author has developed the technique shown in Figure 72, where a summary explanation is shown that reflects the actual delay in the construction portion of the schedule. This quick reference element of the schedule facilitates team awareness of potential progress impediments. Another important distinction is the delay in the owner WBS delay section can include a projected duration of additional delay. The actual activities in the construction WBS section should only include actual delays. With this technique, the owner WBS delay section serves as a proactive forecasting tool which enables more accurate mitigation measures.

At this point in the process, the contractor should refresh their understanding of delays and the appropriate usage.

(a) Concurrent delays are not allowed for submission as a delay. A concurrent delay occurs when a combination of two or more independent delays happen during the same period. Only one of the causes can be submitted as a delay claim and it must impact the critical path of the project.

Owner Scope Modifications		08-Jun-15	08-Jun-15 A
A3410	Unforeseen Underground Debris for Utilities Discove	08-Jun-15	08-Jun-15 A
A3420	Change order submitted to HISD for approval	02-Jul-15	02-Jul-15 A
Time Extensions due to Delays		12-Feb-15	12-Feb-15 A
A3230	Potential Delay for Resevoir Removal	12-Feb-15	12-Feb-15 A
ACT120	Potential Delay for TMC Gate Relocation		
A3240	Potential Delay for Backfill Permit Receipt		

Figure 72: Time Delay WBS Section in Master Project Schedule.

(b) Avoidable delays are not allowed for submission as a delay. This type of delay is also referred to as a nonexcusable delay. This type of delay occurs when the cause is based upon the contractor's actions within their control; that is, low productivity, accidents, etc. In these instances, the contractor is responsible to mitigate the schedule previously approved substantial completion. The contractor is not entitled to a time extension or monetary compensation.

(c) Excusable delays are allowable as a submission for delay and may be approved by the owner. The circumstances involved in these instances are where delays are beyond the contractor's control; that is, owner changes in scope, major weather impacts, etc. In these cases, the contractor is entitled to both time and monetary compensation for the delay. In the instances where the contract prohibits monetary compensation for delays but allows time extension, the delay is referred to as "No Damage for Delay".

While these types of delays are generally described in this text, it is critical that the contractor reviews the terms of the contract with regard to schedule issues. Caveats in the contract may create subtle changes to these basic definitions.

- Step 19 – Do not delete previously approved baseline activities. The reasoning for this mandate is twofold. Deleting activities can drastically alter the linkage and logic from the baseline schedule; therefore, the recommended practice is to:

(d) place this description next to the activity name (VOID) and,
(e) mark the activity as complete so that the logic will calculate properly. It is recommended to use the decision date for voiding the activity as the completion date.

Secondly, by utilizing void decision dates as the activities completion, this provides an audit trail for future reference.

- Step 20 – Cycle time for project updates should be evaluated based upon the complexity and risk associated with the project. Predicated upon the fact that change is the only constant, it is important to assess the cycle time for updates. When the risk or complexity is commensurately greater, it is recommended that the updates become more frequent. This methodology enables the project team to mitigate issues at a faster pace and more proactively. On a dynamic project, flux in the cycle time is an indicator that the project schedule is actively managed. As a secondary recommendation, the author suggests that cycle time never stretch beyond the standard one-month review period.

While Chapter 7 may appear to be a checklist, the primary intent is to convey risk mitigation techniques in the schedule creation and updating process. Risk mitigation applies to all aspects of project time management, not just the actual activities within a schedule. Once this paradigm is adopted by the project team, the learning culture becomes established. Team members begin to "see around corners" anticipating risk factors to the project and the processes required for construction. The risk identification approach creates time to address impeding factors to timely project success. The key to successful project execution is creating the time for managers to problem solve issues before they occur. The steps and sequencing above should assist in managing your day job as a project manager and creating an effective execution plan as a scheduler.

Chapter 8

Did Your Planning Meet the Desired Deadline?

There are two basic constructs that have proven to be consistent in the construction industry since its inception; time and money are success indicators for any project. The scope of this book has revolved around the management and mitigation of time for the purpose of meeting the originally agreed upon substantial completion. The common thread for all the techniques presented has been identification and inclusion of the riskiest elements of a project. The author has postulated that the highest risk variables create the greatest opportunities for both time loss and gain. This reasoning is based upon extensive construction experience and proven techniques. At this point in the process, it is time to ask, "Did your planning meet the desired deadline?" Through the preponderance of evidence, a culture of learning and adaptation is created. Success in any endeavor is measured by how well we learn, grow, and increase performance.

A review of the principles in this book facilitate an effective evaluation of schedule management. A list of key planning and scheduling elements to be included in the lessons learned evaluation are enumerated below.

- Were the key stakeholders of the project surveyed for their high-risk related activities?
- Which methods were employed to generate the list of risk activities?
- Were the high-risk discoveries translated into activities within the schedule?
- Did the critical path include the highest risk activities, regardless of whom was responsible for the task?
- Was a baseline schedule agreed upon between the owner and contractor within the commencement phase of construction?
- Was there a continuous effort to create float in the critical path?
- Was there a regular periodic reporting method employed to communicate the project schedule?
- Was the schedule update report shared with a broad range of project decision-makers?
- Was a shared language developed?

If the answer to each of the questions above was "yes," then the probabilities are high that the team has addressed risk in the project schedule. Suffice to say, if some of the answers were "no," then the team should work to address these issues to ensure a complete risk inclusive execution plan. Proactivity would assert that these questions be addressed as the team is preparing the plan and schedule for

the project, not just when it's
time to complete lessons learned.
Continuous feedback to these
questions during the construc-

A Team Culture of Learning Creates Momentum & Continuous Improvement

tion process will facilitate a risk-centered mentality and increase probabilities for
successful time management.

The proposed methods for discovery and development of risk strategies
include the following tools:

- Sticky note collaborative sessions.
- Application of the risk management matrix during planning.
- Productivity tracking on high-risk operations, and
- Proactive schedule reporting.

In reality, any method for risk identification is acceptable. The primary focal
point is that the team shares the same vision for a success through collaborative
prioritization of risk elements. Since project decision-makers span the range of
both technically savvy construction professionals and nontechnical stakeholders,
it is critical that communication of risk is simplified. A significant attribute of
risk management is the frequency of transmission. Frequency applies to both
reporting and float creation. The professional who possesses the attitude that
time management is an organic process will consistently outperform those who
believe that the baseline schedule is static and nonchangeable. The baseline
schedule represents the original agreement or approach to meeting the contractual time requirements. Change is inevitable, and projects will evolve with their
own rhythm. Based upon this under-

The perfect plan is an illusion, Change is inevitable.

standing, contingency and float creation are essential means to maintain
the original substantial completion
date. Float will "be your friend" by embracing the approach of manifesting positive float early and often throughout the construction execution process.

The most important take away from this manuscript is the mindset of
embracing risk in the design and construction process. The driving factor for
writing this book has been that risk is often discussed but seldom translated into
real activities within construction planning and scheduling. "Hoping for the
best" and ignoring high-risk aspects of the delivery process is folly at best.
Creation of a learning culture where risk is
explored, contingency plans developed to
address the conditions of risk, and inclusion of
risk activities within the schedule are depend-

Float Creation is Your Friend

able processes to increase successful completion. The author's experience has
repeatedly shown that "it's not how you start, but how you finish" that sets the
remembrance of performance. Strong finishes are fueled by the constant drive
for float creation and risk management. Happy risk hunting to you all!

Appendix

Examples of Potential Risk Elements from the Perspective of a Contractor

(1) Temporary Structures
- Installation of temporary structures such as sheet pilings, shoring, cofferdams, dewatering, underpinning
- Installation of scaffolding
- Crane locations, Crane mats, Rigging

(2) Ordinances
- Laws regarding working hours
- Laws regarding noise
- Laws regarding burning
- Laws regarding water pollution
- Laws regarding removal of soils
- Status of adjoining contracts

(3) Transportation
- Water (Barges), railroads, trucking
- Traffic problems − height restriction
- Road − load limits, surface condition, oversize load hours
- Local laws governing movement of equipment

(4) Weather Conditions
- Historical local weather reports
- Anticipated monthly rainfall
- Anticipated monthly temperatures
- Anticipated monthly snowfall
- Anticipated monthly loss work days

(5) Surface Conditions
- Material
- Brush and tree removal
- Effect of rain, snow, tidewater and season of the year
- Depth of topsoil
- Waterways (streams, lakes, ponds, rivers)

(6) Subsurface Conditions
- Material type
- Geological faults − rocks, seams, etc.
- Water level

- Soil erosion control and sedimentation plan
- Installation of permanent structures such as caissons, piling and deep foundations

(7) Access to the Site
- Turning radius
- Union work rules
- Notice to proceed
- Local permits

(8) Adjacent Site Conditions
- Physical conditions as related to shoring and cribbing and underpinning
- Local laws or regulations which will govern the condition in which borrow pits, stock piles, surrounding ground surfaces must be left in original conditions

(9) Temporary Services
- Available power source
- Power panel boxes
- Temporary heat
- Temporary sanitary facilities
- Temporary storage
- Temporary parking
- Temporary fencing
- Job trailers
- Detour routes and signage

(10) Material Availability
- Quantity available
- Quality of the materials
- Distance or location of materials
- Fuel supply

(11) Labor Availability
- Number of craft workers available
- Travel distance
- Work rules
- Labor contracts
- Transportation of workers
- Housing facilities
- Training requirements

(12) Public Safety Requirements
- Need for police, flag, and traffic control
- Need for fencing animals
- Need for shoring of bridges and adjacent structures
- Need for detours
- Laws regarding public safety (State or Federal)
- Occupational Safety and Health Administration (OSHA)
- Hauling conditions
- Grades or elevation changes
- Surface

- Width, height, and weight of load
- Length of haul
- Traffic conditions
- Road and bridge limitations
- Railroad, traffic lights or lift bridges
(13) Owner Requirements
  - Phasing needs
  - Interim milestone dates
  - Owner provided equipment decisions and delivery

Examples of Potential Risk Elements from the Perspective of an Owner

(1) Temporary Structures
  - Are they necessary and does it require relinquishing valuable operational space?
(2) Adjacent Operational Requirements
  - What operational ordinances are at risk due to the construction process? How will construction impact the prime operations of the owner's business?
(3) Transportation
  - Is end user access to the operating facility in jeopardy due to construction?
  - What access measures need to be installed to create safe access by the owner's personnel and visitors?
(4) Environmental Hazards
  - Are there any material discharges from construction operations that pose a safety hazard to the persons on site?
  - Are there any underground hazards on the construction site?
(5) Adjacent Site Conditions
  - Will adjacent businesses be impacted by the construction activities?
  - Have those businesses been notified of the proposed construction activities and measures taken to maintain their normal operations?
(6) Public Safety Requirements
  - Identification of Directional Signage for vehicular traffic changes due to construction operations.
  - Demarcation of Pedestrian Access around the site with protective barriers due to construction operations.
  - Protection from falling building materials onto the general public during construction.
(7) Owner Requirements
  - Phasing needs identified.
  - Interim milestone dates required.
(8) Owner Furnished Equipment
  - Procurement and installation of special equipment which requires hard connections to MEP systems.

(9) Owner Staffing for Activation
  • Is the current operational staff sufficiently prepared to activate and operate the building? If the answer is "no", what resources or processes need to be employed to achieve activation?
  • Is the maintenance staff prepared to activate and operate the building with their current resources? If the answer is "no", what resources or processes need to be employed to achieve activation?

Definitions

Acceleration
Typically a measure to counteract a delay incurred due to the owner's responsibility. This type of plan can result in additional costs either through earlier material delivery or additional manpower. Typically, this type of plan is evaluated by comparing the proposed cost to the operational loss due to the delayed opening of the facility.

Activation
The period in time between substantial completion and building opening for its intended operation. The purpose of this phase is for the end users to occupy the building and test equipment, so the building can operate. The challenge created during this phase is that the majority of people in the building are non-construction professionals who are unfamiliar with the layout of the building. Additionally, construction workers are attempting to complete the punchlist operations. The massive people count can create difficult logistic issues for personnel and furniture migration.

Contingency Plan
It represents a secondary approach to completing a scope of work. The topic of the contingency plan should be carefully considered to include the highest risk elements with the greatest probability of non-planned performance. Knowledge gathering, and creativity are instrumental to the development of an effective plan. In order to maximize the float or "time contingency" in a schedule, it is best to create these plans earlier in the project, rather than later. This proactive approach allows the project team to make more adjustments which increase the probability of success.

Critical Path
The longest path, in terms of time, to complete a project. This path is determined through a forward and backward path algorithm within a computerized software program. The determination of the activities which make up a critical path is based upon relationship logic and activity duration as defined by the contractor's scheduler.

Executive Summary Schedule Report
It is for the purpose of communicating to non-technical stakeholders the status of the project schedule. Conceptually, current and projected risk activities should be addressed with a proposed means to mitigate potential delays. The report should also include delays to date and the current approved substantial completion date. The actual mitigation measures employed and their status toward success should also be included.

Mitigation
The action taken to amend a situation to its original completion date. Based upon this understanding mitigation measures can be components to the contingency plan. This definition implies that mitigation happens after schedule slippage.

Planning
The process of creating a critical path timeline for execution of a project. The intent is to answer the question of when activities must occur in order to meet a proposed deadline for a project.

Scheduling
Scheduling is the successor activity to planning. This process is generally executed by a contractor who is knowledgeable with computer scheduling programs and construction building techniques.

Schedule Contingency (Float)
The time period that can be used for unanticipated delays without impacting the overall substantial completion date. This term is synonymous with the scheduling term of float. Float can be shown in the schedule via two methods: either included within the durations for each activity or as a single activity at the end of the schedule. It is important to evaluate the balance of float at a regular reporting interval, so the team can manage the probability of completion as per the contractual deadline. There are three status indicators of float. Positive float is an indicator that the completion is ahead of the target date. Negative float means the project is behind schedule. Zero float indicates that the project is on schedule. There are two types of float: free and total float. Free float represents the amount of time that a task can slip before it impacts the early start of the successor task. Total float indicates the amount of time that an activity's start can be delayed without affecting the planned project completion date based upon critical path calculations.

Weather Day
An allowable delay to the schedule based upon the contractually agreed upon definition of weather. This is generally specific to a geographical region and may include rain, wind, heat, etc. Refer to the contract as to the

definition specified for your project as each owner approaches this definition in a slightly different manner.

White paper

A report which summarizes a complex issue. The report will typically include pros and cons relative to making a decision on the topic presented.

Work Breakdown Structure

A hierarchy of work that must be completed within the project and groups activities into various phases (e.g.: Foundation, Structure, Finishes, Close-out, etc.). This categorization of activities into smaller construction phases allows identification of similar activity functions. An example would be the exterior skin of a building. While there may be multiple trades involved with completing the skin, the overall process is critical for the major milestone of drying in the building.

References

Analytics, Dodge Data. (2014). The top 400 contractors (2013–2014). Retrieved from http://enr.construction.com/toplists/Top-Contractors/001-100.asp

Analytics, Dodge Data. (2015a). The top 50 program management firms (2015). Retrieved from http://enr.construction.com/toplists/Top-Program-Managers/001-100.asp

Analytics, Dodge Data. (2015b). The top 500 design firms (2014–2015). Retrieved from http://enr.construction.com/toplists/Top?Design?Firms/401?500.asp

Broaddus Associates. (2009). Broaddus-Muñoz named project manager for university hospital expansion. Retrieved from https://broaddusassociates.com/news/item/41-broaddus-munoz-named-project-manager-for-university-hospital-expansionttp://munozandcompany.com/about/

Hospital, Houston Methodist. (2017). Facts & statistics. Retrieved from http://www.houstonmethodist.org/about-us/what-we-believe/facts-statistics/#

Merriam-Webster. (2017a). Definition of mitigation. Retrieved from https://www.merriam-webster.com/dictionary/mitigation

Merriam-Webster. (2017b). Definition of contingency plan. Retrieved from https://www.merriam-webster.com/dictionary/contingencyplan

Pulsinelli, O. (2014). New York-based firm buys Houston's WHR architects. Retrieved from http://www.bizjournals.com/houston/news/2014/10/01/new-york-based-firm-buys-houstons-whr-architects.html?s=print

List of Figures & Tables

Index

Printed in the United States
By Bookmasters